18 QUESTIONS
ABOUT LIFE AND T

Nottingham University Press
Manor Farm, Main Street, Thrumpton
Nottingham, NG11 0AX, United Kingdom
www.nup.com

NOTTINGHAM

First published 2011

British Library Cataloguing in Publication Data

18 Questions about Life and the Universe
P Altman

ISBN 978-1-908062-56-7

Disclaimer

Cover images: Top left - Mercy from Wikimedia Commons; Middle left - DadosMuchasCaras from Wikimedia Commons

Typeset by Nottingham University Press, Nottingham
Printed and bound by Lightning Source, UK

18 Questions about Life and the Universe

Peter Altman

For Joan, for all your support and encouragement

Contents

Author's Note

The self-imposed rules that govern the content of this book are that answers are based, as far as possible, on scientific evidence. Any opinions are entirely my own and although I have tried to base them on established facts and principles, some people may disagree with my conclusions. That's fine, as it leads to interesting debates and discussions.

The main section of the book, Part Two, consists of 18 chapters, each of which is formed as a question. The chapters end in one of two ways: **THIS IS AN ANSWERABLE QUESTION** or **THIS IS AN UNANSWERABLE QUESTION.** I consider a question to be answerable if, in my opinion, there is sufficient good evidence, or lack of it, to provide an answer. I consider a question to be unanswerable if, again in my opinion, there is insufficient good evidence to provide an answer. In these cases I have provided what I call a Best Guess answer.

Scientific facts presented as such are, to the best of my knowledge, correct and in line with current thinking. Any errors are solely my responsibility.

Clearly, each of the chapters could be expanded to fill a book on its own but my purpose has been to cover many questions rather than to delve very deeply into just a few. Those interested in finding out more about a specific topic will find many more specialised books and articles at their disposal.

Acknowledgements

I would like to thank Dr Clifford Adams, Managing Editor at Nottingham University Press, for publishing my book. It's a hard task for an unknown author, and I am pleased that he noticed my tee shirt at the London Book Fair when I passed by the NUP stand.

© Helsingin Sanomat 2011

I would also like to thank his colleagues Sarah Keeling, Production Manager, and Rosalyn Webb, Business Development Manager, for their friendly and efficient handling of my material, and for producing a fine-looking book.

I am grateful to various friends and colleagues for reading early manuscript drafts and for their helpful comments, several of which I have incorporated into the book. Finally, I have attempted to contact copyright holders where illustrations are either not my own or where I believe them not to be in the public domain, and I would like to thank all those who have kindly given permission for the use of their material in this book. The Publishers and I apologise if we have unwittingly infringed any copyrights; such oversights will be corrected in any future printings if we are made aware of them.

PREFACE

Have you ever wondered how the Universe started and how it might end, or whether there is alien life on other planets, or whether time travel is possible, whether there is a Creator, how Life began, whether coincidences have any meaning, or whether we will ever know everything? There must be answers to these and other fascinating questions about Life and the Universe but at the moment we do not know what these answers are. However, that doesn't mean that we can't make a best guess based on what we do know.

That's what this book is about. It's a journey through the Universe which allows us to wonder about the answers to many questions. Some we can answer with a fair degree of confidence but for others we can only give a best guess.

For example, 50 or more years ago, people might have gazed up at the Moon and wondered what the far side looked like (since the same part is always facing the Earth). This would have been an unanswerable question. We can now give an answer to this question – pretty much like the near side - because in 1959 the Soviet space probe Luna 3 photographed it for the first time. Before 1959 however, all we could have done was to make a best guess.

Fig. 1 shows the most detailed image yet of the moon's far side, photographed by NASA's Lunar Reconnaissance Orbiter spacecraft in 2010.

Fig. 1 Far side of the Moon

This is a good example of how new knowledge answered a question that was previously unanswerable.

When asked a question such as "*Do you think that there is intelligent life on other planets?*" most people would say that with so many billions of stars and planets in the Universe, life must surely have arisen somewhere else apart from Earth. As we'll see in Question 8 however, although this answer might seem reasonable it isn't really justifiable on current evidence.

This book is a journey from the beginning of time through to a far distant future that attempts to provide answers to a number of questions based on current knowledge. Some of the questions can be answered with a good degree of certainty; other answers are more speculative and can only be a best guess. One day, hopefully, better answers will be available.

We need to begin though with a bit of physics...

PART ONE

Why do things happen the way they do?

A HOW LONG IS A PIECE OF STRING? WHY THE PHYSICAL CONSTANTS ARE THE VALUES THAT THEY ARE.

This well-known question is usually given as an answer to some other unanswerable question, such as what will the weather be like on this day next year? In other words, there is no answer since the string can obviously be any length you choose just as the weather is unknowable so far into the future.

But what if we ask a different question, such as how long is a ruler? Most people would give the answer 12 inches or 1 foot.

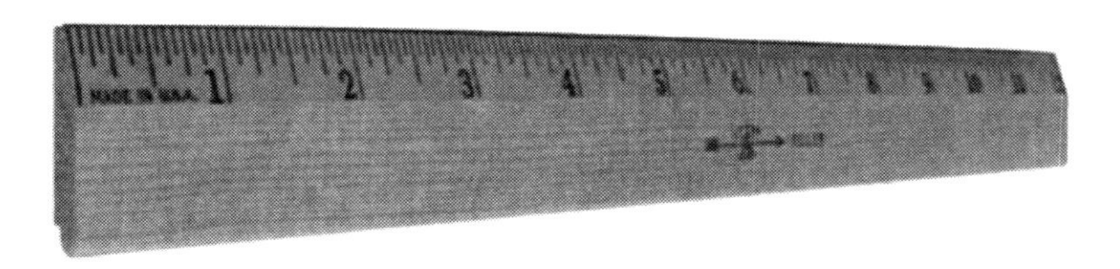

Fig. 2 A 12 inch ruler

It's true that most rulers are indeed 12 inches long although you can buy shorter and longer ones. This is just convention – 12 inches is a convenient length that can be carried about or stored in a drawer or case but a shorter one would be easier to carry in a pocket and a longer one would be more accurate for making longer measurements.

So unlike the piece of string question, where we cannot give an answer at all, we can at least say that most rulers are 12 inches long.

Now let's consider a question like what is the speed of light? For many thousands of years it wasn't known whether light travelled instantaneously or just very quickly, but in 1676, the Danish astronomer Ole Rømer proved that light had a finite speed. It took another 300 years for the actual speed to be measured with an accuracy of 1 in 250 million so that in 1975 the speed of light through a vacuum was known to be 299,792,458 metres per second or 186,282 miles 698 yards 2 feet and $5^{21}/_{127}$ inches per second.

In 1983, the metre was redefined as the distance travelled by light in a vacuum in $^{1}/_{299,792,458}$ of a second. This then created an absolute value for the speed of light with no margin of error at all. So, when we ask what is the speed of light in a vacuum, we can give an absolutely correct answer - 299,792,458 metres per second, or the miles per second figure given above.

The speed of light through a vacuum is one of many numbers in science known as *constants*. The name says it all; these values never change unlike the length of a ruler or the length of a piece of string or the height of a man or the weight of an orange. These are known as *variables* since their value can and does vary.

Other examples of constants are the acceleration of a falling object due to gravity, Planck's constant which deals with the size of packets of energy in quantum mechanics, and the elementary charge which is the value of the electric charge on a single proton or electron. There are many other such constants. They may not appear to impact very much on everyday life but they are very important nevertheless.

The Universe obeys the laws of physics and these laws are defined by the values of the physical constants. What's really interesting though is that the Universe would be a very different place if the physical constants had had different values. Small differences of just a few percent in some of the constants would affect the properties of the elements and, as a consequence, all sorts of other things such as star formation, the reactions going on inside stars, the formation of heavy elements, the retention of planetary atmospheres, the properties of water, and many other processes that are all necessary for the formation of the Universe we know and the creation and sustainability of life.

Let's try and put this into the context of everyday life. Imagine a train as in Fig. 3. It runs on rails and goes where the rails go until it gets to a set of points where it can go one of several ways depending on how the points are set. The majority of the World's railways use the *Standard Gauge* for their track. This means that the distance between the inside edges of the track is exactly 1,435 mm, or 4 ft 8½ in.

Now imagine that the train is approaching a set of points where it is due to take a left fork. However, the track to the left is of the 1,422 mm (4ft 8in) gauge. What do you think is going to happen to the 100 tonne locomotive as it careers round the bend?

Fig. 3 Train on standard gauge (4 ft 8½ in) track

Its wheels are now too far apart to fit onto the track so it will jump the rails to be followed by all the coaches behind. If it hasn't fallen over by now it will do very soon resulting in a spectacular train crash.

This is an example of how a very small change in the size of something (less than 1% in this particular case) can have a very large effect on a wider scale. In a similar way, very small changes in the size of the physical constants would have a very large effect on the structure of the Universe.

All life on Earth is based on the element *carbon*. This is because carbon has a unique property among the elements in that it can form very long and stable chains and other shapes of molecules. Silicon is an alternative element that has been proposed for an alien biochemistry since it can also form long chains although it does not have the versatility of carbon in its reactions with other elements.

The chemical properties of carbon enable it to form the vast variety of chemicals necessary for life to exist, such as proteins, carbohydrates, fats, DNA, vitamins, hormones, to give just a few examples. DNA, for example, is a very long molecule because it has to carry a great deal of information which then needs to be copied very accurately.

Lots of small molecules would not be a substitute since then there would be no information about how these are ordered, and the order is vital.

This is easily demonstrated as follows. My name is Peter Altman. These 11 letters (or 12 characters if you count the space between the two names), in their specific order, spell out my name. They can be re-arranged in many different orders, such as for example this one – Pamela Trent. The letters are the same but the order is different and it no longer spells out my name. I'm not Pamela Trent.

Another example of the importance of ordered information is your credit card's PIN. You must enter the digits in the correct order for the PIN to be recognised – just using the correct digits isn't enough. If the number is 1234, then entering 4321 won't work.

It's the same with DNA. Using the name analogy, the order of the smaller molecules (the letters in my name) that make up the large DNA structure (my complete name) is important since if the order is different, then so is the information carried by the DNA.

As mentioned above, the properties of the elements are dependent on the values of the various physical constants. Had these been different, then carbon would not have this unique property, and either we'd have no life at all, or maybe we'd have life forms based on another element. Silicon maybe, or phosphorus, or sulphur.

Why then do the constants have the values that they do? That's a fascinating question since if they didn't, we probably wouldn't be here to ask the question. Physicists and astronomers call it the *Fine Tuned Universe.*

There's no good answer to this question. Some would say that a Creator designed it this way and others would not accept the concept of the fine tuned Universe at all. They would say that this is just one of many Universes which have existed or co-exist or could exist and it's the one that happens to have the values that permit Life and an organised Universe to develop.

Think of a pack of cards. The cards can be arranged in 52! different ways (factorial 52, which equals 52 x 51 x 50 x.....3 x 2 x 1). This is a truly massive number equal to about 10^{68}. It's so large that it's almost certain that every time you shuffle a pack of cards, the order of the cards that results has never been seen before and will never be seen again.

One possible result is that the cards are back in their original factory order, each in their own suits numerically arranged from Ace downwards. Similarly, one possible Universe out of a huge number of possible Universes is the one that has the values of the constants that are compatible with Life as we know it. We are in that one because it's the only one we could be in.

Here's another way of looking at it.

Fig. 4 Flower pot universes

Fig. 4 shows 5 flower pots. Each one is filled with a different substance as follows:

pot 1	concrete
pot 2	glue
pot 3	acid
pot 4	soil
pot 5	glass

Imagine though that there are billions of flower pots each filled with different substances but only pot 4 contains something in which plants can grow. The plants in pot 4 might think, if they could do so, *'Hey! Aren't we lucky that our pot contains soil so that we can grow. If it had contained something else then we wouldn't be here.'* That's certainly true but luck has nothing to do with it. With billions of pots collectively containing every possible substance, the plants *had* to grow in pot 4 because that's the only pot in which they *could* grow.

It's the same with our Universe. The soil it contains, in our case the values of the physical constants that govern the laws of physics and chemistry, make it possible for Life as we know it to be here. Billions of other Universes might exist which contain different substances, that is, values of physical constants, that make it impossible for our type of Life to exist.

So we're not *lucky* that the constants are the values that permit Life to exist just like the plants in pot 4 aren't *lucky* to be in that pot; it's the only pot they could be in and this is the only Universe that we could be in.

This is known as the *anthropic principle.* It's really a philosophical argument that for the Universe to be observed by living beings, its conditions must be such that the living beings that do the observing can exist. Even though this may seem complicated on first reading it's actually very simple.

The fact that we're here means that the conditions in which we live are such that we can be here.

It is of course also possible that we might be able to exist in another Universe with slightly different values for the physical constants. They can't be too different though since calculations have shown that differences of just a few percentage points are enough to make a very different Universe, and one in which our type of Life just couldn't exist.

As stated earlier in the chapter, life as we know it is based on carbon. This is because carbon has the unique property amongst the elements of being able to form molecules with very long chains and other shapes such as rings. It is therefore ideally suited as the basis of the vast variety of different molecules required by living creatures.

All the carbon and other elements in the Universe (apart from hydrogen and helium which were created during the Big Bang) were produced by chemical reactions inside stars. When these stars came to the end of their life cycles they sometimes exploded into super novae and their contents, including the carbon, were sent out into space during the explosion. This debris eventually coalesced into planets, and this is where the carbon in our bodies came from.

Now here's the important bit. Computer calculations have shown that if the values of a few of the physical constants had been different by only a few percentage points, then the reactions that produce the carbon inside stars couldn't have taken place. No carbon; no carbon-based life; no us.

We've seen the example of the train leaving its tracks in Fig. 3. Here is another example. A few years ago, on holiday in La Paz, Bolivia, a friend asked for a boiled egg for breakfast. She knew how she liked them cooked so asked for a 3 minute egg. The waiter complied and brought her egg which, when cracked open, was uncooked and unedible.

La Paz is at an altitude of about 4,000 metres above sea level. Under these conditions, the atmospheric pressure is considerably lower than at sea level and one of the consequences is that water boils at about 86°C rather than at 100°C. This is not hot enough to boil an egg in 3 minutes. No wonder the waiter was smiling.

This is just a rather trivial example of how a change in the value of a physical constant – the boiling point of water – can affect the outcome of another process – the boiling of an egg.

The constants are the values that they are. Had they been different, then yes, the Universe would also have been different and so would we, if we were there at all, but so what?

We aren't the purpose of the Universe, we are merely one of its products. (See Question 2 for a more detailed discussion about the possibility of other Universes – the *multiverse* concept).

This chapter's message

The Universe is defined by the laws of physics and these laws operate the way they do because of the numerical values of a range of numbers known as constants. Had these numbers been different then the Universe would have been different and so would life. Does this mean that the values of the constants were designed to be the values that they are?

Some would say so but there is a counter argument. The constants are the values that they are. If they had been different then so would the Universe and so would we but so what? What we have and see now is the product of the laws of physics as they happen to be: had they been different then the Universe and everything in it would also have been different and someone or something else in a different type of Universe might be asking this question.

Cause always come before effect. You don't scream out in pain (effect) before you hit your thumb with a hammer (cause); a balloon doesn't burst (effect) before you prick it with a pin (cause); an object doesn't fall to the ground (effect) before you let go of it (cause).

The way in which the Universe and everything in it, including life, has developed is the effect; the laws of physics and the values of the constants are the cause. If the cause had been different then so would have the effects.

B THE LAWS OF PHYSICS

The laws of physics (sometimes referred to as the laws of nature) are dependent on the values of the constants discussed in the previous chapter, and these laws in turn are responsible for how and why things happen the way they do. Before we can begin to provide answers to our questions, we need to consider this with a few examples.

We know what we know as a result of observations and experiments performed by many people over many years. These observations and experiments resulted in discoveries which were then confirmed by other people repeating them and getting the same answers. After many such repeats, we can safely assume that the answer is true.

GRAVITY

If you drop something, it falls. If you were to do it a thousand times, it would drop every time. If a thousand people in a thousand different places were to do, it would drop every time. So it's safe to conclude that things fall if you drop them.

Now let's add to the experiment by dropping objects of different weights at the same time.

We would then discover, after dropping lots of different objects, that they all take the same time to reach the ground, irrespective of their weight. That's quite a surprising discovery, since instinctively we would expect a heavier object to fall faster.

There's an apocryphal story about Galileo, the 16th Century Italian physicist and mathematician. Galileo was born in Pisa, and the story tells that he dropped two objects of different weights from the top of the Leaning Tower of Pisa.

The idea behind this experiment was to disprove the theory of Aristotle, the Greek philosopher who lived 2,000 years earlier in the 4th Century BCE, who said that heavier objects fall to the ground faster.

Galileo's experiment proved that Aristotle was wrong (although you would have thought that Aristotle would have tried it out before announcing an incorrect result to the World).

Disregarding the complication of air resistance on very light objects such as pieces of paper, the attraction due to gravity is independent of the weight of the object. More can then be learnt about gravity by dropping objects from different heights and timing the fall to the ground. You would also find out that objects speed up as they fall, that is, they accelerate. This is known as the acceleration due to gravity and has the symbol g.

Main point: dropped objects fall to the ground. No-one *makes* them fall, they just do because of the existing laws of physics.

Salt mountain

Take a piece of paper and make a small hole in the centre, then fold it a few times through the hole to make a sort of cone shape when you hold it. Pour some salt on to the paper and hold it above a table or other flat surface. Let the salt fall through the hole and try not to move the paper while this is happening.

This is the sort of thing you'll see (Figs. 5 and 6).

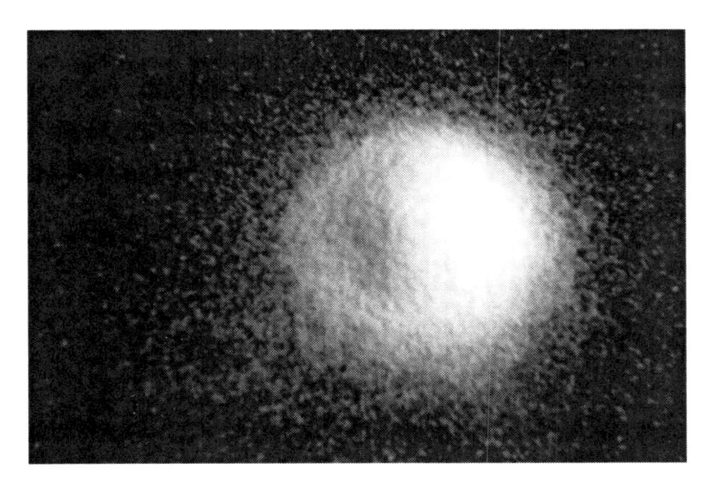

Fig. 5 Top view of salt mountain

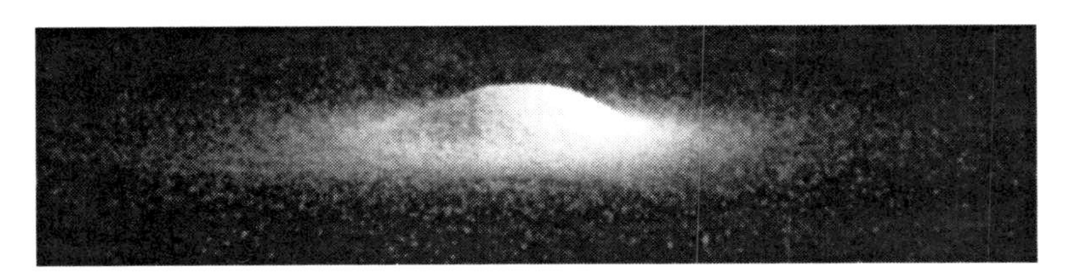

Fig. 6 Side view of salt mountain

The salt has fallen into an almost perfect circular shape and has formed a symmetrical mound that gradually falls away the further it gets from the centre. Why? Why hasn't it formed a tall column of salt, or made a triangular shape on the table?

Every time you do this you'll get the same result. If a million people do it, they'll all get the same result.

Why – because it's physics. Gravity makes the salt fall through the hole. The salt comes to rest in a state of equilibrium where it is stable, and a column of salt crystals would not be stable enough to survive for more than a brief fraction of a second. The round shape is due to the equal forces on the salt from all directions.

Main point: Salt dropped through a small hole on to a flat surface comes to rest in a symmetrical and circular heap. No-one makes it form such a shape, it just does because of the existing laws of physics.

Black and White: Hot and Cold

In the days before sports sponsorship, people who played tennis and cricket and other active sports tended to wear white clothing. Why? Something looks white because all the light hitting it bounces back into your eye, white light being a composite of all the spectral colours red, orange, yellow, green, blue, indigo and violet. The infra-red wavelengths, invisible to the human eye but perceived as heat, also bounce back. Because of this the white garment remains cool.

Conversely, something looks black because all the light hitting it is absorbed and none is reflected back into your eyes, black being an absence of all colour. This time, the infra-red heat rays are also absorbed so the black garment gets warm.

So that's why white, or light, colours keep you cool, and black or dark colours make you warm. White cars are cooler in the summer than black cars, and black cars are warmer in the winter than white cars.

Look for a house painted black and white as in Fig. 7.

Fig. 7 House painted black and white

If it's a hot sunny day, feel the black paint and the white paint, and you'll notice that the black areas feel much warmer than the white ones. The black paint will also tend to peel more quickly since it will be absorbing heat energy much more efficiently than the white paint, and this will eventually make it crackle and dry up and peel off.

Main point: a black colour absorbs heat and gets warm; a white colour reflects heat and stays cool. No-one makes this happen, it just does because of the existing laws of physics.

Spectrum

If light passes through a prism – a piece of cut glass – or raindrops or a window at a certain angle, it will be split into 7 separate colours, known as the visible spectrum. The colours are red, orange, yellow, green, blue, indigo and violet.

These 7 colours always appear and always in this order. Green and orange never change places, for example, nor do any of the other colours. Why? Light consists of waves, and different colours have different wavelengths. Red light has the longest wavelength (the distance between similar points from one wave to the next). Each colour from red, through orange, yellow, green, blue, indigo and violet, has a shorter wavelength than the previous one, ending up at violet which has the shortest. Since each colour has a different wavelength, it is perceived differently by the eye, which sees it as a different colour. The point though is that there is a progression of wavelengths from long to short, and they always go in order, which is why the colours always appear in the same order.

Main point: White light can be split up into 7 colours by passing it through a prism. No-one makes it do this, it just does because of the existing laws of physcs.

Round the Bend

Sometimes, when approaching a bend in the road, there will be a lower speed limit posted. The road may also be banked, that is, built up to slope inwards.

Why? Think of an object, maybe a conker, tied to a piece of string. Swing the string in a circle. What happens? The string gets taut and the conker goes round and round. If the conker is only loosely tied to the string, it will come away and continue going until it hits something and falls to the ground.

In other words, the conker is trying to get away – it gets as far as it can and the string gets taut. If it can overcome the knot it will fly off.

This is called *centrifugal force.* You can experience it yourself on a merry-go-round at the funfair. As it goes round, your body will tend to move outwards away from the centre and you will feel pressure as you press against the back of the seat. This effect becomes more pronounced the faster you go round.

Now back to the car on the bend. As you drive round the bend, centrifugal force tends to force the car outwards, and off the road. Obviously this must be avoided, so a maximum speed is posted. Banking the road inwards lessens the effect of the force and therefore lessens the risk of running off the road (Fig. 8).

When the road is designed, the engineers need to work out the maximum safe speed and the angle of banking. They use the following formula:

$$\tan\theta = v^2 / rg$$

tan is a trigonometric function; θ is the angle of the banking; **v** is the maximum speed before the car would leave the road; **r** is the radius of the bend; and **g** is the acceleration due to gravity.

The formula is given here not in the expectation that it will be meaningful to anyone who is not conversant with applied mathematics, but to illustrate the point that physical laws are exact and reproducible, and can be expressed in precise mathematical terms.

Fig. 8 Driving round a bend

Main point: If an object goes round a bend faster than a certain speed, it will leave its original path and continue in a straight line until it either gradually falls to the ground or until it hits an object in its way. No-one makes it leave its original path, it just does because of the existing laws of physics.

Π (PI)

The ratio of the circumference of a circle to its diameter is known by the Greek letter π, pronounced pi (Fig. 9). This name was first proposed by a Welsh mathematician William Jones in 1706. He chose the letter pi, which is a Greek *P,* from the initial letter of the word *perimeter.*

pi is often written as equal to 22 / 7, or 3.142, but these are just approximations. It can't be written down as an exact number, whether decimal or fraction. To date, it has been calculated by computer to one trillion decimal places!

pi had the same value when the ancient Egyptians tried to calculate it 5,000 years ago; it has the same value on the Moon or on Mars or indeed anywhere in the Universe. Why is it the value it is? It just is.

It is one of a range of numbers that mathematicians and physicists call *constants* (see Part One A). As the word implies, these are numbers that are believed to always have been and always will be the same.

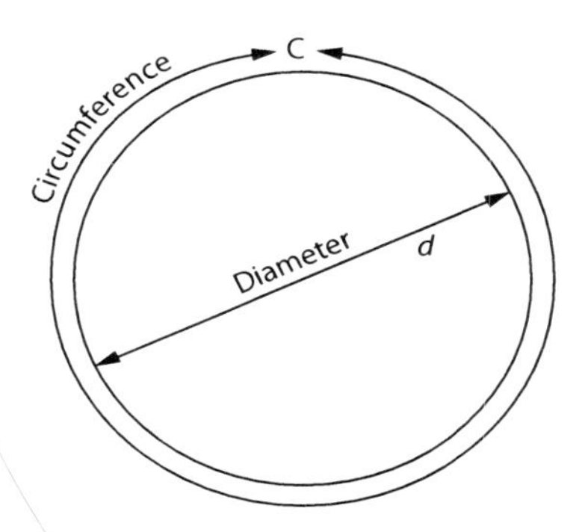

Fig. 9 Circle showing the circumference C and the diameter d. C divided by d = pi

Main point: pi never changes its value. No-one made it this value, it just is because of the existing laws of mathematics.

The Tangent and the Chord

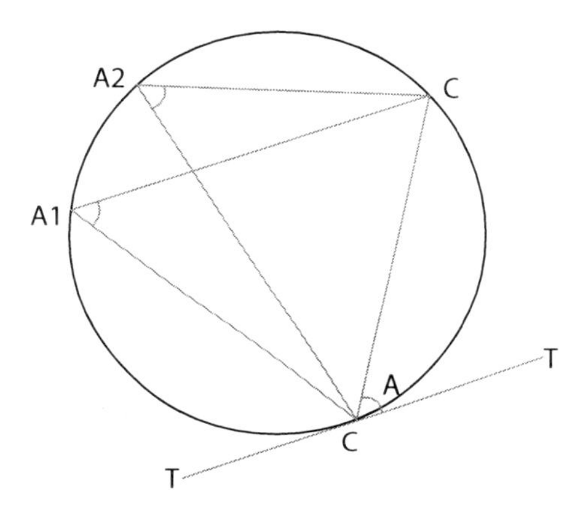

Fig. 10 The tangent and the chord theorem

Fig. 10 shows a circle with some lines. The line from **T** to **T** is called a *tangent* and only touches the circle once, in this case at the point **C**. The line from this point **C** to the second point **C** is called a *chord*. The angle between the tangent **TT** and the chord **CC** has been labelled **A**. It's called, obviously, **the angle between the tangent and the chord**.

Taking the chord **CC** as the baseline, it's easy to construct a triangle by drawing one line from each end of the line **CC** to the circumference of the circle.

In this example it has been done twice, so we have two triangles which both have the chord **CC** as their base. The top angles of these triangles are labelled **A1** and **A2**.

It's clear that many triangles could be drawn in this way, each with an angle at the top where the point of the triangle touches the circle.

Now for the clever bit. The angle between the tangent and the chord, **A**, is always exactly the same as the top angle of the triangle, in our example **A1** and **A2**. Mathematically it's expressed as:

The angle between the tangent and the chord is equal to the angle in the alternate segment.

It doesn't matter how big or small the circle is, or where you draw the tangent and the chord, or how many triangles you draw, or whether you do it in Manchester or on Mars, it will always be the case that **the angle between the tangent and the chord is equal to the angle in the alternate segment**. This is just one of numerous mathematical facts that are due to the inherent properties of geometrical shapes.

Main point: Shapes, and numbers, as well as objects, have inherent properties that make them behave in certain ways. No-one makes this happen, it just does because of the existing laws of mathematics.

The Speed of Light

The speed of light (and of radio and other electromagnetic waves) in a vacuum is equal to 186,282 miles (or 299,792,458 metres) per second. Why? Why doesn't it go a bit faster or a bit slower? No reason; that's just the speed it is.

And it's the same everywhere, whether it's coming from the Sun or the Moon or the stars or your table lamp (but then with due allowance for its passage through the air which will slow it down by about 0.03%).

Main point: Electromagnetic waves (light; radio; X ray; gamma etc) travel at 186,282 miles per second in a vacuum. No-one makes them travel at this speed, they just do because of the existing laws of physics.

Eclipses

Partial eclipses of the Sun and Moon are common but the really spectacular event is a total eclipse of the Sun (Fig. 11). This was last seen in the United Kingdom in August 1999. Using astronomical and mathematical data, it's been possible to work out the dates of many total eclipses that took place hundreds and even thousands of years ago.

This has made it possible to place accurate dates on contemporary records and paintings which describe such an event.

Fig. 11 Total solar eclipse (from Wikipedia Commons - Luc Viatour)

It is also possible to calculate the times and dates of eclipses that haven't happened yet and this has been done for many years into the future.

Main point: The fact that it is possible to calculate the exact date and time of a future eclipse, and then to confirm that it does actually occur at this exact date and time, proves that the laws of physics and mathematics are true and unchanging.

Quartz Crystals

Some quartz crystals are shown in Fig. 12.

Fig. 12 Quartz crystals

Quartz is a very common mineral in the Earth's crust. Chemically, it is silicon dioxide, and it forms crystals in the shape of 6-sided prisms. Why does it do this? The properties of the silicon and oxygen atoms are such that they combine in a certain shape and this in turn gives rise to the characteristic shape of the quartz crystal.

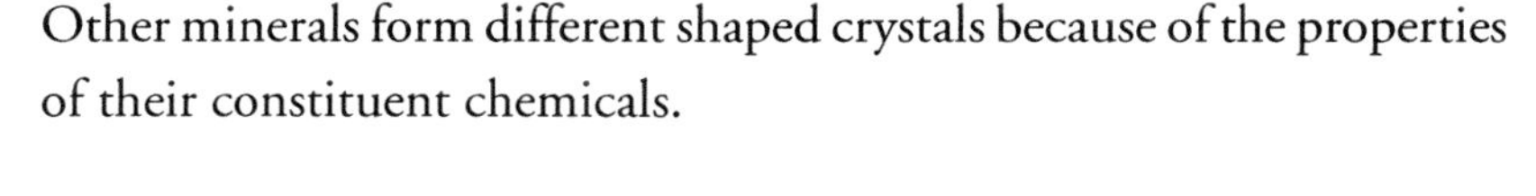

Other minerals form different shaped crystals because of the properties of their constituent chemicals.

Main point: No-one makes the quartz crystals the shape that they are. They just are because of the existing laws of physics and chemistry.

Undiscovered Laws

What about as yet undiscovered laws? Could there be some laws of physics that we don't know about? Yes, of course there could and there almost certainly are.

However, these are likely to be at the very extremes of nature, such as at very low or very high temperatures or pressures, or involving very small or very large objects. These are all situations that are difficult to study. It is most unlikely that any as yet undiscovered laws of physics would impact on everyday situations. Why? Because we'd know about them already from our everyday experiences as detailed above.

We've seen that things happen in predictable ways. If, for example, there was an unknown law about how centrifugal force works, then this could obviously not be taken into account when surveyors and engineers work out how to bank a bending road. The calculated maximum speed for the bend of, say, 50 mph, would then be wrong, and cars would keep going off the road. The fact that this doesn't happen means that for all practical purposes, we do know all about centrifugal force.

Another example is the notion of a twin Earth planet on the other side of the Sun which is forever hidden from our view. This is an old idea dating back many thousands of years, and even spawned a 1950s American Science Fiction comic, *Twin Earths* (Fig. 13).

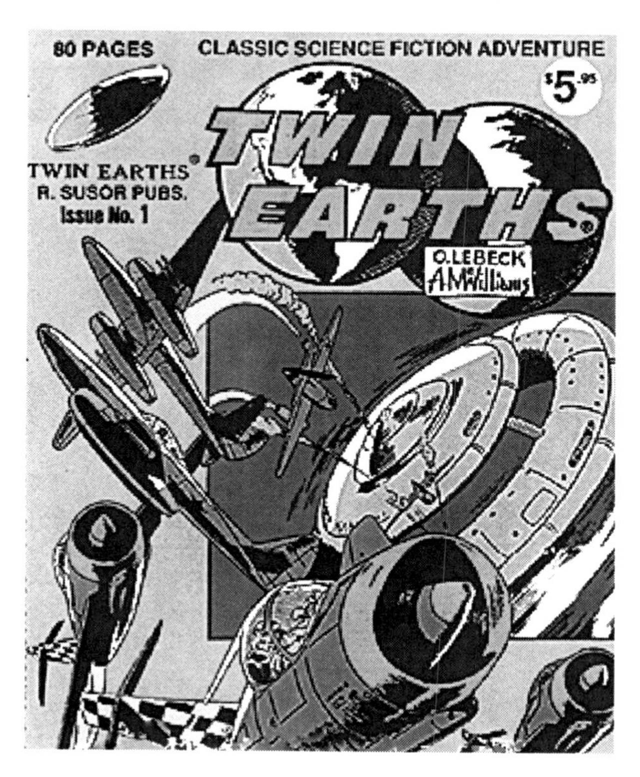

Fig. 13 Twin Earths comic cover; issue number 1
© R. Susor Publications

We knew that such a planet didn't exist long before spacecraft confirmed it visually. How? Because if it was there, then the calculations made by astronomers for the motions of the other planets would be wrong since any gravitational effects of the putative planet would not have been taken into account.

This example leads us into an interesting discussion and shows how an undiscovered law of physics was eventually found.

It all started with the planet Mercury. Observations of Mercury's orbit round the Sun never quite matched up with predictions based on calculations of the orbit. In 1840, the French mathematician and astronomer Urbain Le Verrier (Fig. 14) was given the task of trying to find out why. Nearly 20 years later, Le Verrier published an extremely thorough study of Mercury's orbit based on numerous meticulous observations.

Fig. 14 Urbain Le Verrier (1811 – 1877) Wikimedia Commons

There was still a difference between the observed and calculated orbits, albeit a very small one. However, what is considered small in everyday life - a train arriving 5 minutes late for example - is a big deal in astronomical terms. The precision of Le Verrier's work led to the inescapable conclusion that there was indeed an unknown factor at work. What might this unknown factor be? Since the other planets behaved as predicted, this unknown factor would seem to be peculiar to Mercury.

In 1846, Le Verrier had been involved in the discovery of the planet Neptune. Uranus, the 7th planet out from the Sun, was the first 'new' planet to be discovered, the other six having been known from antiquity. William Herschel, a British astronomer, discovered the planet on 13 March 1781 and received world-wide acclaim. Many astronomers then attempted to calculate the orbit of the new planet but it did not behave as they predicted.

Some people suggested that Newton's Laws of Motion didn't apply so far away from the Sun (nearly 2 billion miles). Others suggested that perhaps the orbit was being affected by the gravitational pull of yet another unknown planet even further away.

Calculations were made as to where such a planet should be in order to have such an effect, and on 23 September 1846, the planet Neptune was discovered almost exactly where it was predicted. Various astronomers share the credit for this – Le Verrier who did the maths and the German astronomers Johann Galle and Heinrich d'Arrest who looked for and found the planet where Le Verrier predicted it would be.

Further studies of the Uranian and Neptunian orbits suggested that there might be a further unknown planet outside the orbit of Neptune. This eventually led to the discovery of Pluto by the American astronomer Clyde Tombaugh on 18 February 1930.

In view of his success with discovering Neptune, it is perhaps not surprising that Le Verrier suggested that the discrepancy in Mercury's orbit was due to the existence of an unknown planet between Mercury and the Sun. He proposed the name *Vulcan* for this planet, Vulcan being the Roman god of volcanoes and fire. This was thought to be an appropriate name for an object so close to the Sun. (The fictional planet Vulcan of *Star Trek* fame was supposedly in orbit around the star 40 Eridani, 16 light years from Earth).

Given Le Verrier's reputation as a planet finder, astronomers around the World started looking for this new planet but without success. Le Verrier died in 1877 still convinced that he had discovered a second planet. However, since close to 20 years of observations by astronomers in many countries had failed to find the new planet, the consensus opinion was that it did not exist.

If Vulcan did not exist, how then could the discrepancies in Mercury's orbit be explained? The answer did not come for another 40 years when, in 1915, Albert Einstein published his work on the Theory of Relativity.

Einstein provided a new approach to the understanding of gravity as compared to the Newtonian Laws of Motion. The apparent discrepancies in the orbit of Mercury were fully explained by the application of Einsteinian rather than Newtonian mechanics and the need for an extra planet disappeared.

Relativity also resulted in changes to the calculated orbits of the other planets but since these were much further from the Sun than Mercury, the differences between the observed and calculated orbits were so very small that they were within accepted margins of error. In 1915, therefore, a new Law of Physics was discovered which explained observations that previously were unexplainable.

Main point: It's most unlikely that there are any unknown Laws of Physics still to be discovered that will have a significant effect on how most things behave. It's only at the very extremes of nature – the unbelievably small, large, hot or cold, that we're likely to discover something really new.

SUMMARY

There are many other examples that could be given but these will suffice.

What do these events have in common? They always happen and the result is always the same. If you drop an object, it will always fall to the ground. If you let salt run through a small hole, it will always form a symmetrical heap.

White objects will always feel cool and black objects will always feel warm when the Sun shines on them. Light always splits up into 7 colours of the spectrum in the same order. If you divide the circumference of a circle by its diameter, the answer will always be 3.1415926..., known as *pi*. Quartz always forms characteristically shaped crystals, as do many other minerals.

The timing of eclipses follows strict rules which can be predicted accurately by astronomers. Electromagnetic waves always travel at 186,282 miles per second through a vacuum.

These are the Laws of Physics (and Mathematics). They apply everywhere and at all times; London and Lisbon, Manchester and the Moon, Monday morning and Friday afternoon, 1 billion years ago and today. By studying stars and galaxies at different distances from the Earth, ranging from a few light years to billions of light years, it's possible to confirm that the fundamental laws of physics are the same now as they were billions of years ago (more of this in Question 4).

So what? Well, these examples prove that Nature works by obeying certain fundamental rules and laws which are always and have always and will always be true. No-one *makes* them happen or be as they are, they just do or just are due to the values of the various physical constants.

This chapter's message

Light waves, cars, black paint, circles, table salt, planets and all other things have inherent properties that make them behave in a certain way under certain conditions. No-one ***makes*** them behave in this way - they just do, and always have and always will because of the values of the physical constants. It's what scientists refer to as the Laws of Physics. There may be more laws waiting to be discovered but these won't change what we currently observe although they may help with some explanations of things that we currently can't explain.

C Evidence and how to assess it

The remaining chapters in this book each consider a specific question and end either with an answer or, if no answer is available, a *Best Guess* answer based on current knowledge. This knowledge comes from scientific evidence, so it is worth discussing briefly the nature of evidence and when we should believe it and when we shouldn't.

Evidence is information used to determine whether something is true or not true. It is an essential component in all walks of life.

Let's consider a few examples.

1. You pay for something in a shop and receive your change. You complain that the change is incorrect since you gave the shopkeeper a £50 note which you recall had a stain on it and a corner torn off. The shopkeeper insists you gave him a £10 note. On examination, the cash tray is found to contain a £50 note just as you described. The shopkeeper says that the cash tray had previously contained another £50 note with a stain and its corner torn off but although this is theoretically possible, most people would discount this as extremely unlikely. By far the most reasonable conclusion therefore is that you were right and the shopkeeper was wrong.

Extremely unlikey

2. Professor Christiaan Barnard, the South African surgeon, performed the world's first human to human heart transplant in 1967. Although it is theoretically possible that another surgeon had performed this procedure at an earlier date, there are no reports of such an event.

   Since it was such a pioneering procedure, it seems extremely unlikely that an earlier transplant would have received no publicity. So if someone claims to have pre-dated Barnard's achievement but provides no supporting evidence, most people would discount the claim

3. If you look at the sky and observe the motion of the Sun during the course of a day, you would notice that it rises in the east, travels across the sky and sets in the west. The obvious conclusion from this evidence is that the Sun moves around the Earth, resulting in Ptolemy's second century *geocentric* model of The Solar System with the Sun (and planets) all revolving round a stationary Earth.

   Although there had been earlier suggestions of a *heliocentric* solar system, it wasn't until 1543 that Copernicus published his work showing that the Sun was at the centre of the solar system with all the planets moving round it. Such an arrangement was the only one which complied with all the observations of not only the Sun but of the planets as well.

   So here we can see that an apparently obvious conclusion can turn out to be false once additional evidence (the motions of the planets) is considered.

4. You have decided to visit a clairvoyant on the recommendation of a friend. You have recently lost your mother and you would like to contact her. After the visit your friend asks you how you got on. "*She was fantastic. She knew that my mother had died and told me that she still loved me and thought about me, and told me to look after the kids and never forget her.*"

   Perhaps you shouldn't be so easily impressed. You would have looked a little sad as you were thinking about your recently deceased mother, so a guess at a lost parent, perhaps after one or two preliminary questions, isn't so remarkable.

   Based on your own apparent age, the clairvoyant took a shot that you had already lost your father – they tend to die first – and so guessed it was your mother. The rest of the 'reading' is pure generality and would apply to practically everyone in your position.

   There are two ways in which the clairvoyant could have gathered the knowledge required to present your reading (although very little actual 'knowledge' was needed). Either she really did have psychic powers enabling her to make contact with dead people, or she combined a couple of educated guesses with some generalities. There is a name for this process – it's called *Cold Reading*. Skillfully peformed, it can achieve seemingly remarkable results.

   Be reasonable; which one sounds more likely?

   These examples show that it is still possible to mount a defence against seemingly irrefutable evidence, but the stronger the original evidence the more fanciful the defence needs to be.

Who would believe that the shopkeeper already had a £50 note with a stain and corner torn off in his cash desk? It is possible of course but so unlikely that very few people would believe him.

Who would believe that a surgeon had performed a human to human heart transplant before 3 December 1967, kept the procedure a secret and had no supporting documentation or other evidence? Again, it's possible but so unlikely that very few people would believe him.

The third example about the Sun and solar system shows that incomplete evidence can lead to false conclusions which in this case had persisted for well over 1,000 years.

The fourth example of the clairvoyant shows that, faced with two explanations, many people will go for the most outrageous if that is the one they *want* to believe. Their desire to believe overrides their common sense.

It's like the story of the Cottingley fairies. These were a series of photographs taken between 1917 and 1920 by two young schoolgirls who lived in Cottingley in Yorkshire.

Frances Griffiths (10) and her cousin Elsie Wright (16) took the photographs as a joke. They cut some pictures of fairies out of a book, propped them up in their garden with hatpins, and posed with them for a series of photographs (Fig. 15).

Although never intended as a deception, the photographs became public and sparked a great deal of interest. Sir Arthur Conan

Doyle, who was a spiritualist and believed in psychic phenomena, was convinced that the photographs were genuine as did other prominent people of the time. It wasn't until many years later, in 1983, that Griffiths and Wright admitted that the photographs had been faked.

Fig. 15 The first of several 'fairy' photographs (1917). This one shows Frances Griffiths. © National Media Museum/SSPL

It seems bizarre that two children could create a series of rather crude fake photographs and fool such eminent people as Sir Arthur Conan Doyle. The reason was that Conan Doyle and the others *wanted* to believe. This desire was so strong that it became the overriding factor in their decision.

It is important therefore to consider evidence on its own merits and not be swayed by what you would want the answer to be.

Evidence forms an essential component of law and of science. It rarely provides an absolute truth since there is usually an alternative explanation and it is up to the observer to balance the probabilities and come to a conclusion.

5. Imagine that you are a member of a jury. You are shown a photograph of the suspect taken by a CCTV camera outside a house that was burgled. Of course, many innocent people would have walked past the same location but you take one look at the photo and straight away make a presumption of guilt. Why? Because he *looks* like a villain: dour expression, unshaven, hooded. But that's not evidence – that's prejudice.

Evidence would be his fingerprints on the stolen items or shards of glass from the windows on his coat. That's the way verdicts are arrived at in courts of law and the same procedure should be adopted here when we come to consider the questions posed in this book. Preconceived ideas and wishful thinking have no place; it's the evidence that counts. And if there is no evidence? Innocent until proved guilty. If there's no evidence for X then we have to say that there is no X.

We should mention the concept of *Ockham's Razor.* William of Ockham (also sometimes spelt as Occam) was a 14th Century English friar and philosopher who developed the concept that the explanation of any event should make as few assumptions as possible. In other words, the simplest explanation is likely to be the best one, as in our final example.

6. Let's imagine that you are a police officer and you come across a road traffic accident shortly after the event. Two cars, say a red one and a black one, have collided head on in the left hand lane of a busy road. The red car is on the wrong side of the road and likely to be at fault. The driver of this car admits that he is on the wrong side but says that it is not his fault, even though he was making a telephone call at the time.

   His reason was that a sudden very strong wind blew his car to the other side of the road.

   Most people would not believe the story of the sudden wind and would say that the accident was the red car driver's fault because he lost concentration while on his mobile telephone. Unknowingly therefore, the principle of Ockham's Razor has been applied.

   We don't need to consider a sudden strong wind to explain what happened. Yes it's possible but the simpler explanation is the more likely.

It will be obvious that some questions are easier to answer than others. This has nothing to do with a particular question being 'hard', but more to do with whether the information to answer it is available.

Consider this question. *Who was the first person to run a mile in under 4 minutes?* Answer: Sir Roger Bannister. Is it possible that someone had run a sub four minute mile before Bannister but that this has remained a secret for over 50 years? Like Professor Barnard's heart transplant, theoretically yes, but very few people would take that view. This is therefore an easy question to answer, either through personal knowledge or through a bit of research.

Now consider this question. *Which city will host the 2020 Olympic Games?* There are currently over 30 candidate cities so this question cannot be answered at present no matter how much research is done. The information to answer it just isn't available. All we can do is to make a best guess based on such information as happens to be available about likely venues.

An introductory chapter in the next section explains a little more about how the questions have been categorised into *answerable* and *unanswerable.*

This chapter's message

When deciding whether something is true or not, one considers the evidence. Sometimes it is clear cut and there is little if any doubt about the answer, and sometimes one has to make choice based on one's interpretation of the evidence and the balance of probabilities.

Even though it may be hard to do, one shouldn't let bias and pre-conceived ideas get in the way of an impartial consideration of the available facts.

Some answers are easier to arrive at than others, depending on the available evidence. Good evidence gives us confidence in our answer and poor or little evidence makes the answer more of a guess. In some cases though, that's the best we can do.

PART TWO

The Questions

Introduction

Answerable and unanswerable questions

Questions are often categorised according to their degree of difficulty. Here are two that most people would call easy:

What is the capital of France? Answer: Paris

What is the square root of 100? Answer: 10

Conversely, here are two questions that most people would call hard:

What is the telephone dialling code for Bolivia? Answer: 591

What is the currency unit of Guatemala? Answer: quetzal

The distinction between easy and hard is of course subjective. Any question to which you know the answer would, for you, be easy, and vice versa. However, the four sample questions here all have one thing in common, and that is they all do have a definitive answer. You may know the answer but if you don't, you can look it up because an answer does exist. They are all answerable questions.

Now contrast this with these questions:

On which date was Jesus Christ born?

Will there be another civil supersonic aeroplane?

There are no definitive answers to these questions. One might be able to give an educated guess based on some research but that is about the best one could do. The first question deals with a past event so although there once was an answer, the information is no longer available so no accurate answer can now be given. The second question deals with a future event so there can be no answer, just a best guess.

For the purposes of this book and the questions that it poses, we shall categorise the questions as follows. An **Answerable** question is one which, based on available information, is answerable with a reasonable degree of certainty. An **Unanswerable** question is one which, even with the benefit of available information, is still unanswerable with a reasonable degree of certainty. In these cases all one can do is to make a best guess based on whatever information happens to be available.

Question 1:
How did the Universe begin?

We live on a planet (Earth) which circles a star (the Sun) which is part of a galaxy (the Milky Way).

Fig. 16 Drawing of the Milky Way as viewed from outside the galaxy showing the position of the Sun as a white dot. © NASA

Fig. 16 is a drawing of what our galaxy might look like if viewed from outside the galaxy. Our Sun is located near one of the arms known as the *Local Spur*, and is marked with a dot. Galaxies are huge objects. Ours is about 100,000 light years wide (that means it would take light 100,000 years to go from one side to the other, equivalent to about 600,000, million million miles), and about 1,000 light years thick, and contains perhaps 500 billion stars.

Fig. 17 The Milky Way as seen on a clear night © NASA

When you look up at the sky on a clear night away from bright city lights, you might see a band of hazy light stretching across the sky (Fig. 17). This is one of the spiral arms of the Milky Way galaxy.

Some galaxies have neighbours (in the astronomical sense, since they can be millions of light years away) and such a group of galaxies is called a *cluster*.

Some clusters can even form super-clusters. Some galaxies are spiral, like ours, and others can be elliptical or other shapes. Current estimates put the total number of galaxies in the Universe at about 100 billion. The space between galaxies, the inter-galactic space, is pretty empty with only a few atoms drifting about and also very cold at a temperature of about minus 270°C. The oldest stars and galaxies are over 13 billion years old as measured by a variety of astronomical techniques. The observable Universe is estimated to be about 95 billion light years in diameter.

This brief description is given in attempt to create some impression of the size and scope of our Universe.

Objects such as the Andromeda nebula are just about visible to the naked eye and have been known since antiquity. Many other nebulae had been discovered with the aid of the telescope, and up to the 1920s, it was thought that all these objects were located within our own Milky Way galaxy.

In 1924, Edwin Hubble (Fig. 18; 1889-1953), an American astronomer, used a newly discovered method for measuring stellar distances and calculated that the Andromeda nebula was over 1 million light years away (the currently accepted value is about 2.5 million but he had made his point). Since the diameter of the Milky Way galaxy is about 100,000 light years, Andromeda must lie outside our galaxy.

Fig. 18 Edwin Hubble. Wikimedia Commons.

Hubble had therefore proved that Andromeda and all the other nebulae lay far outside our own galaxy and that they were separate galaxies in their own right.

Hubble also showed that these other galaxies were receding from us in all directions, giving rise to the conclusion that the Universe was expanding. The further the galaxies were from us, the faster they were receding.

This made it possible to calculate backwards to the time when the Universe started to expand, and this figure would represent the age of the Universe. Current estimates give this age as close to 13.7 billion years.

This implies that about 13.7 billion years ago the Universe consisted of a single point that, for some reason, became unstable and exploded into all the matter that we see today. What this single point was, how and why it exploded, are questions we can't answer except to say that, as described by the various examples in Part One, given a certain set of circumstances it *had* to happen. No-one made it happen, it just did because of the laws of physics.

It may have been very unlikely but that doesn't matter since once was enough.

This is known as the *Big Bang Theory* of the origin of the Universe. It was developed in the late 1920s by Georges Lemaître, a Roman Catholic priest. Although there are a number of theories about the mechanism of creating something from nothing, none of them provide an answer that is satisfactory in layman's terms as to how this could come about.

A *theory* is an attempt to explain something that has happened which can then be tested by experiment. For example, if a light suddenly goes out, we might expect to find a broken filament if we examine the bulb. In the same way, many scientific theories enable predictions to be made which can then be tested. If the predictions turn out to be true, then this lends support to the correctness of the theory. Experimental support for the Big Bang Theory came in 1992 as a result of NASA's COBE (**CO**smic **B**ackground **E**xplorer) satellite.

If you look at the sky with an optical telescope, the space between heavenly objects – planets, stars and galaxies – is black. However, if you look with a radio telescope, then you can detect a faint glow. This glow is not due to any planets, stars or galaxies and is therefore called *background* radiation. It is strongest in the *microwave* region of the electromagnetic spectrum. Because of this, it is known as the Cosmic Microwave Background (CMB) radiation.

The existence of CMB was predicted in 1948 and discovered mostly by accident in 1964 by two American radio astronomers, Arno Penzias and Robert Wilson. It earned them the Nobel Prize for Physics in 1978.

When NASA's COBE satellite made very detailed measurements of the CMB, it found small temperature variations in what was originally thought to be a completely uniform glow. These variations were visualised as a temperature map of the sky and published in 1992 (Fig. 19). The variations were small – of the order of one part in 100,000 – but they were there.

This map was published in newspapers and magazines around the World. Why was it so significant? Because the variations in the CMB, shown as different colours in the map, are precisely what would be expected if the Universe began as the result of a Big Bang. No other theory apart from the Big Bang theory has yet been proposed that can explain these variations.

Fig. 19 Variations in the Cosmic Background Radiation, shown as different colours (shades of grey here), as detected and measured by the COBE satellite.
Wikimedia Commons

Current knowledge therefore leads us to suppose that the Universe did indeed begin its life as the result of a Big Bang.

Having said that, there is increasing interest amongst cosmologists to provide an answer to the obvious next question – what happened *before* the Big Bang? (Fig. 20).

Fig 20. Pondering the mathematics of the Big Bang

We'll come back to this again in a moment. Meanwhile though we shall consider our options about the origin of the Universe. We have three.

1. The Universe arose on its own (Big Bang or other process)

2. The Universe has always existed

3. The Universe was made by a Creator

1 THE AROSE ON ITS OWN OPTION

If the Universe arose on its own, how did this happen? We don't know, but there are some clues from things that we do know.

We've seen already that atoms and molecules, and light and salt, and cars and black paint, have certain inherent properties which means that they behave in a certain way. Under the same conditions, they will always behave in this way (see Part One – Why things happen the way they do). So let's imagine a situation just before the Universe came into being.

There's nothing and then suddenly there's something. How come? Something happened, we don't know what, but it resulted in the appearance of all the material that we see in the Universe today. Try and relate this to, for example, visible light being split into the 7 colours of the spectrum after it passes through a prism. First there's just white light and then suddenly there are 7 colours because something happened.

Or your car goes round a bend too fast and runs off the road. You're driving along and suddenly, because something happens (going too fast), you're off the road. These changes are due to the inherent properties of light and the car.

It's hard to imagine nothing and something happening to it and then there's something, but this is really in the realms of theoretical physics and higher mathematics so is always going to be difficult for the non-specialist to visualise.

However, we need to remember the lessons of Part One which showed us that all things have inherent properties that make them behave in a certain way under certain conditions, due to the values of the physical constants. Whatever the conditions were at the moment the Universe was created, it would have been inevitable that the Universe appeared due to the inherent properties of whatever existed at that moment, even if in layman's terms, it was 'nothing'.

We also need to remember Einstein's famous equation $E = mc^2$ (E = energy; m = mass; c = the speed of light). This tells us that mass and energy are interchangeable so that an amount of energy could have been turned into an amount of mass.

But then where did the energy come from? A good question with no satisfactory answer at present, even amongst cosmologists (Fig. 20). But that doesn't mean there isn't one – it just means we don't know what it is.

Now let's consider the second option.

2 The ALWAYS EXISTED option

Saying that the Universe has always existed immediately removes the problem of how it started. It didn't start, it's always existed. This is still consistent with the Big Bang theory and just implies that all the matter that existed in the point that exploded at the Big Bang – known as a *singularity* – has always existed. The Universe may be an oscillating system going from Big Bang to Big Crunch every trillion years or so, repeating the process for ever (see Question 15. It's always existed and always will exist.

The concept of something having always existed is hard to visualise. We can't ask *'Where did it come from?'* because it didn't come from anywhere – it's just always been there although not necessarily in the same form.

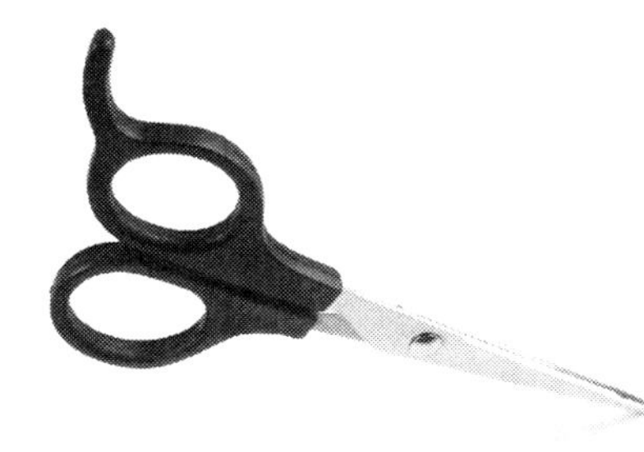

What happens if we ask where, say, a pair of scissors came from.

Here is a flow chart showing the origin of the scissors.

office drawer > shop > wholesaler > manufacturer > sheet metal supplier > metal refiner > mining company > iron ore > Earth > super-nova > Milky Way galaxy > primordial dust cloud > Universe > singularity > ? or previous Big Crunch

We could take *any* object and create a similar flow chart which would always end at the singularity, because that is the ultimate source object from which the Universe arose. Up to this point, our two options are the same. The AROSE ON ITS OWN option ends at the singularity because we have no knowledge of what happened before that.

The ALWAYS EXISTED option doesn't end at all and merely continues backwards into a previous Big Crunch.

The concept of something having always existed is not an easy one to visualise or to accept, and it certainly won't be an easy one to test or to prove. We have to include it though, until such time as there is evidence one way or the other.

Finally, we have the Creator option.

3 THE CREATOR OPTION

If you think about it, what does it actually mean to say that the Universe arose through the work of a Creator or designer, or in other words, God? Like the concept of life arising not on Earth but extra-terrestrially and brought here by a meteorite or comet or asteroid (see Question 3), it doesn't solve the problem but just places it somewhere else.

If the Universe did result from the work of a Creator, how was it done? There would have to have been a mechanism. Saying that humans can't or aren't meant to understand it doesn't help and just pushes the question away.

So whether the Universe arose by itself as a result of the laws of physics, or has always existed, or arose as a result of the work of a Creator, we still have the same difficulty of how it was done. Invoking a Creator or designer doesn't help with solving this problem. So why bother?

The Universe exists. We don't know whether it came into being at the moment of the Big Bang or whether it has always existed. What we do know is that it is here now. We can at least all agree on that, whatever our faith or belief. Some of us say a Creator did it; some say it happened on its own due to the laws of physics because under a certain situation it *had* to happen (like the car *has* to run off the road if it goes too fast round a bend – Part One B); and some say that it has always existed.

You could argue that a Creator, if he exists, actually had nothing to do. The Universe as it exists today obeys the laws of physics. Planets go round stars in predictable orbits, eclipses happen as predicted, comets appear as predicted, cars going too fast run off the road as predicted, objects fall to the ground as predicted, etc.

Everything that's ever been studied obeys the fundamental laws of nature. Obviously, we can't say that those laws were exactly the same 13.7 billion years ago at the moment of the Big Bang as they are now since no-one was there to investigate them. But even if they were different, it seems reasonable to think that there were laws that governed how matter behaved.

So if the Universe, as a finished article, obeys the laws of physics, then it's reasonable to assume that the Universe, as a not-yet existing article but an about-to-appear article, also obeyed the laws of physics even if those laws weren't the same as they are now.

In other words, as we've said before, under the right conditions, it *had* to arise. A Creator, if he was there at all, had nothing to do except to watch and admire.

So how did this actually work? If it arose rather than always having existed, how could *something* arise out of *nothing*?

Let's digress slightly before we try and address this question.

Ideas about the structure of matter have undergone many changes over time. The early concepts were that everything was made of just 4 elements – *Earth, Air, Fire* and *Water.*

Much later, in the 18th Century, chemists began to realise that these were not elements at all, and started to isolate and list what we now recognise as elements, such as *hydrogen, oxygen, phosphorus, sulphur* etc.

Then, at the end of the 19th Century, it was found that elements could still be sub-divided into even smaller particles, the *electrons, protons* and *neutrons* that make up each atom.

The development of powerful particle accelerators in the 1950s led to the discovery of a huge number of even smaller particles which spawned the entirely new subject of particle physics. *Quarks, mesons, baryons, hadrons,* and *bosons* are just some of the entities that have been discovered. And so it goes on.

It is difficult if not impossible for a layman to form a mental image of what these latest particles 'look' like or are made of. It was easy enough for the atom – a sort of mini solar system with a central nucleus and electrons spinning round it (see Fig. 30). But how do you picture a *boson*?

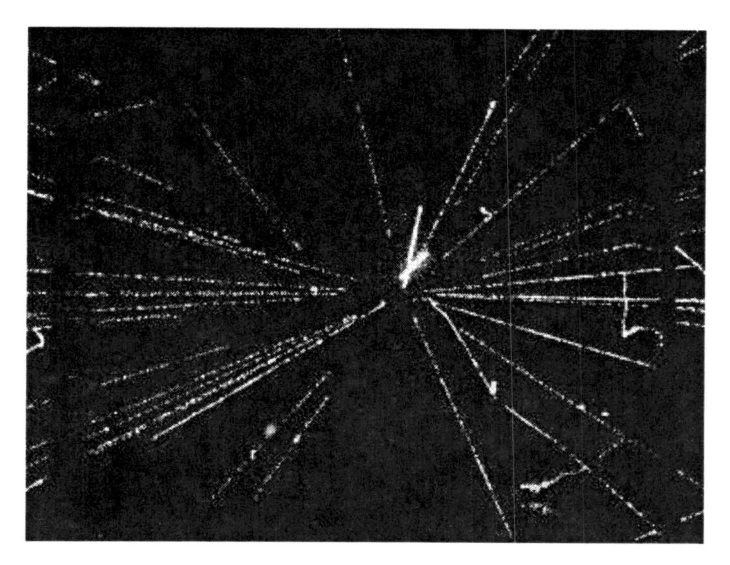

Fig. 21 Spark chamber showing particle collisions (from Wikimedia Commons)

Fig. 21 is about the best we can do. This is an image from a *Spark Chamber*, a device used to visualise what happens when particles are made to collide with each other at high speeds and high energy.

The collision is seen at the centre of the image and the tracks of the resulting smaller particles are seen to be radiating outwards. The angle and energy of these new particles can then be used to characterise them.

So although we can't actually *see* the particles, we can at least see their tracks, so we know they exist.

In addition to all these new particles, there are also 4 basic forces that need to be considered. These are *Gravity*, a long range force that acts as a attraction on all matter; *Electromagnetic Force*, a long range force that acts on particles that have an electric charge; *Strong Nuclear Force,* the force that holds the protons and neutrons together inside the nucleus of an atom; and *Weak Nuclear Force,* the force that causes atoms to disintegrate as in radioactivity.

Scientists from Einstein onwards have long been searching for a Theory of Everything – a Unified Theory that brings together all that is known about the many particles that make up matter and the four forces just described. It's become a Holy Grail of physics.

In the 1980s, a mathematical model called *String Theory* was developed. Briefly, mathematical equations were constructed that appeared to show how all the known particles and forces could be combined and described by one-dimensional *strings* – small units that had only length and no height or width. To make it even more difficult to visualise, these *strings* existed in 11 dimensions! This became known as *M Theory* and it is currently thought by some scientists to represent the elusive Unified Theory of Everything. It also, apparently, allows for *something* to be created out of *nothing* although the explanation of how this could happen is firmly in the realms of theoretical physics and advanced mathematics, and would certainly be unconvincing to anyone not immersed in these fields of research.

Some people might find this explanation unsatisfactory. Phrasing the question at its most basic level, how could all of the vast amount of material in the Universe have arisen from 'nothing'?

Look at it this way. Just because we don't have an answer to a question doesn't mean that there isn't one; it just means that we don't know what it is.

We can illustrate this point with the description of a card trick. The magician asks you to choose a card and then tells you what it is. How does he do it? Here are some possibilities.

All the cards in the pack are the same

Possible, but not in this case because he gives you the pack and you examine them and see that they are all different.

The magician makes you take the card he wants

Possible, but not in this case because he gives you the entire pack and you take the card yourself.

The magician glimpses the card as you take it

Possible, but not in this case because you place the pack under the table, cut the pack wherever you like, take the top card and leave it on your lap under the table and out of everyone's sight.

The cards are marked

Possible, but not in this case because even if they were, the magician never sees the card you take.

The magician changes your card for one in the pack by sleight of hand

Possible, but not in this case because the magician never touches any of the cards.

Some other method that you don't know about

Since you can see the trick performed (Fig. 22; search YouTube for *The Unexplained Card Trick* by FPAltman) you know that it can be done. You just don't know how. Did he read your mind? Did he fortell the future? Or was it something else? It doesn't matter – the point is that something has happened but you don't know how it happened.

Fig. 22 The Unexplained Card Trick

Let's now make a big leap backwards in time to the origin of the Universe. The Universe exists because we can see it and are part of it so if it arose rather than having always existd, it must have come from somewhere. Maybe it was M Theory. Maybe it was something else. We just don't know. Can you see the parallel with the card trick? There's a huge difference in scale of course but the principle is the same. Something has happened but we don't understand how it happened. That's puzzling to be sure but it doesn't mean that there isn't a logical explanation.

Let's summarise these two happenings.

CARD TRICK	*ORIGIN OF THE UNIVERSE*
Can be seen on YouTube	Can be seen all around us
There has to be a method	There has to be a method
Explanation known to magicians	Explanation currently unknown

How many people seeing the card trick would think that the magician had divine help? Not many probably. It's just a trick; I don't know how it's done and it's very clever but it's just a trick.

Absolutely right! So when we come to the problem of how the Universe originated, why can't we adopt a similar attitude? I don't know how it arose but it's here so it had to come from somewhere and one day we'll work it out.

Just because you don't understand how the card trick was done doesn't mean that there isn't a logical explanation. You just haven't been told what it is.

Just because we don't understand how the Universe came into being doesn't mean that there isn't a logical explanation. We just haven't worked it out yet.

Or there is the option of saying that the Universe has always existed thus removing the difficult problem of how it arose out of nothing. It didn't arise; it's always been there. This creates its own difficulty though since it is hard to get one's mind around the concept of something having existed for ever.

✗ THIS IS AN UNANSWERABLE QUESTION ✗

BEST GUESS ANSWER

The Universe exists so either it had to arise or it had to always have existed. It obeys the laws of physics now so it's a reasonable assumption that it has always done so, even before it appeared (although the laws it obeyed then might not be the ones we observe today). Under the right conditions, it *had* to appear (just like, under the right conditions, a car *has* to run off the road – Fig. 9). Introducing an additional process, a Creator, is unnecessary. Applying Ockham's Razor, we can ignore it.

If it has always existed, then there is no need to consider how it arose because it didn't have to – it's always been here.

Our Best Guess answer therefore is actually two answers since at the moment we can't say that one is better than the other. The Universe either arose on its own as a result of the laws of physics and the values of the physical constants by a mechanism currently unknown but without intervention from a creator or designer, or it has always existed.

Question 2:
Are there other Universes?

In PART ONE A, we discussed the intriguing question of why the physical constants have the values that they do. It is well known that if these values were only a little bit different, then life, at least our type of life, could not have developed. This has convinced many people that the Universe must have been designed the way it is, otherwise it does seem like a very fortunate coincidence that the laws of physics are what they are and we can be here.

As with most things however, there are other explanations. Like the flower growing in pot 4 (Fig. 4), this is the only Universe that we could be in, so if it was a different Universe then we wouldn't be here. The Universe came first, and it was OK for us, so we are here.

Even so, it does seem remarkable that the Universe has the laws that make it possible for us to exist. But it's only remarkable if ours is the *only* Universe.

Imagine that you have placed a bet in roulette on number 36 and it comes up (Fig. 24).

That's a 1 in 38 chance (counting the zero and the double zero) and you would be

Fig. 24 Winning number at roulette

surprised and pleased. Now imagine that there are hundreds of roulette wheels in the room and that you have placed a bet on number 36 on every one. It's practically certain that 36 would come up on one or more wheels.

It's the same with the concept of other Universes. This has been called the *Multiverse* theory. Just like the roulette wheels, it's practically certain that, given enough Universes, one or more would have the values of the physical constants necessary for life.

This concept of multiple Universes is not easy to accept or visualise. Where are they? Can we see them? How many are there? Where do they come from? Are they here now?

We can't really answer these questions. All we can say is that if other Universes do exist, then they will be extremely far away so that we could never see them, otherwise we would already have detected signs of their presence. This is known as the *event horizon*, a theoretical boundary beyond which we can never venture because it is so far away that light and other radiation will never reach us no matter how long we wait. As to how many other Universes may exist, we obviously cannot answer this either.

There would however need to be vast numbers of them so that like the roulette wheel analogy, there is a certainty that at least one of them will have laws of physics compatible with our type of life. Some cosmological theories predict that there could be as many as 10^{500} such Universes. This is an unimaginably vast number, easily large enough to allow for enough Universes with different laws so that one or more will have laws compatible with life. Where do they come from? We can perhaps imagine lots of Big Bang moments, each creating its own Universe. Are they here now? We don't know. They may co-exist or they may exist one after the other.

If we wish to try and form a mental image of these Universes we could perhaps imagine a number of floating balloons (Fig. 25).

Fig. 25 Balloons representing separate Universes
(credit: creativedoxfoto/FreeDigitalPhotos.net)

Each is different but unlike the picture they should be imagined as being separated by vast distances with no communication possible between them. They may co-exist or they may exist one after the other. There is no way to know.

The multiverse concept is an interesting idea but not testable with current technology. Some cosmological equations predict that such universes should exist but that's far removed from being proof.

✗ This is an Unanswerable Question ✗

BEST GUESS ANSWER

We have to try and explain why the laws of physics are what they are, thereby enabling life to flourish on a stable planet circling a stable star in a stable galaxy. Small differences in the laws would have made this impossible.

There is a reason the laws are what they are. We can consider some possibilities as to why this is the case.

1. The laws had to be something and they just happened to be right for life to exist.

2. The laws were fine tuned by a Creator so that life could exist.

3. The laws are different in different Universes of which there are so many that at least one had to have the laws needed for life to exist, and that's the one we are in.

There is no evidence that makes one of these possibilities better than the others so at our present level of understanding we cannot make a choice between them. Until such time as we have the necessary evidence, it comes down to what we, as individuals, wish to believe.

QUESTION 3:
HOW DID LIFE BEGIN?

Life is all around us and within us, so we can be sure that it exists. There was a time when it didn't. Depending on your beliefs, this may have been thousands or billions of years ago but for now, this doesn't matter. What does matter is that at some time it didn't exist and then it did. So it had to arise somehow. We can all agree on that.

What are the possibilities? Let's start with these.

It arose on Earth
It came from beyond the Earth

It has been suggested by some people that life didn't arise on Earth at all but was brought here in a primitive form by a meteorite or asteroid collision, and then developed here. This is an interesting theory (known as *exogenesis* – literally *outside origin*) but extremely difficult to prove either way. Also, it doesn't help, even if it were true, because then we have to ask how did it arise in its original location. All this does is to place the origin of life somewhere else; it tells us nothing about *how* it arose.

It's interesting to note however, that many of the building blocks of proteins, known as *amino acids*, have been found in comets and meteorites, suggesting that these compounds, which are essential for life, are quite common in space.

There's another interesting thing. Compounds found in life, such as proteins and carbohydrates, are complex organic molecules. These molecules can exist in two forms, called right-handed and left-handed. A good analogy is a pair of gloves. Like the molecules, they are mirror images of each other. This is known as *chirality*.

For example, we can have **d**-glucose (**d** for dextro, which means right) and **l**-glucose (**l** for laevo, which means left). It turns out that every glucose and other sugar-type molecule found in living creatures is of the **d** type. Conversely, all the amino-acids (building blocks of proteins) found in living creatures are of the **l** form. If such a compound were to be made synthetically in a laboratory, then it would consist of an equal amount of both the **d** and the **l** forms. So how did these two forms become separated and only one went on to become the chosen one for building a living creature? We don't know. And what's really interesting is that when these compounds were discovered in some meteorites and comets, as stated above, it was found that there was more of the **d** sugars than of the **l** sugars, and more of the **l** amino acids than the **d** amino acids. (This is discussed again in regard to another topic in Question 7).

This in itself doesn't prove that life came from space since there could be many reasons why right-handed sugars and left-handed amino acids arose on Earth. Cosmic rays, polarised light, lightning strikes, could all have had an influence on how these compounds arose in the first place. But it's interesting nevertheless.

Current geological estimates give an Earth age of about 4.5 billion years (see Question 4 for more on this). The earliest life forms are thought to have arisen about 3.5 billion years ago, leaving about 1 billion years for the process to get going. We have no way of knowing whether that's fast or slow since there are no life forms from other planets that we know about for us to study and compare.

Let's now return to the question of how life might have arisen. In the absence of any real evidence to the contrary, let's assume that this took place on Earth.

So what are our options? There are only two.

1 Came into being on its own

2 Was the work of a Creator or designer, that is, God

1 The 'ON ITS OWN' option

Someone who believes in a Creator or designer will say that this is impossible, pointing out that an object such as a house or a car or an aeroplane is far less complex than a living creature, and no-one would suggest that houses, cars, and aeroplanes could just appear on their own. So how could life?

Well, on that basis it couldn't. But this is a fallacious argument since the analogy is wrong. Obviously a house could never just appear on its own. You could wait forever for this to happen, but no-one is suggesting this. A house couldn't appear on its own, but what about a brick? - a random piece of rock that happens to have become shaped in such a way that it could fit next to a similar brick. A few bricks like this and you are beginning to get a wall. (This example is given only to illustrate the point of a gradual development rather than of the finished article arriving in one step. It's not meant to suggest that a house could eventually be constructed in this way, which it couldn't).

The mistake that many people make when considering and then rejecting the possibility of the emergence of life on its own is to assume that the finished article is formed straightaway. There's nothing and then suddenly there's a crocodile.

Obviously that's not going to happen. It's a gradual process where each step has to be self-sufficient and to have an advantage over previous steps. Thought of in this way it becomes feasible.

In 1952, two American scientists, Harold Miller and Stanley Urey conducted an experiment. They sealed some of the chemicals thought to be present in the atmosphere of the Earth before life arose (methane, ammonia, water, and hydrogen) in a sterile flask and passed sparks of electricity to simulate lightning through it for a week. Amazingly, when the brew was analysed, it contained many organic compounds including amino acids, the building blocks of proteins, as well as sugars. Similar laboratory experiments by other scientists showed that components of nucleic acids, the building blocks of DNA, could also be produced in this sort of way.

What this meant was that starting only with some of the simple chemicals thought to be present on Earth before life began and passing electric sparks through them, it's possible to create many of the more complex compounds needed for life. It's a bit like the brick – it's not a house but it could be the beginnings of a wall. Once the basic building blocks exist, it's at least possible to imagine them coming together to form more complex molecules and eventually small single cells.

You may wonder how and why an electric current could produce sugars and amino acids and nucleic acids. Remember the salt mountain experiment described in Part ONE – why does the salt form a circle and a symmetrical heap? It just does, due partly to the properties of the salt crystals and partly due to the properties of its surroundings – gravity and the flat table.

Why does your car want to leave the road if you drive too fast round a bend? Centrifugal force, or, if you prefer, the properties of bodies in motion and the influence of gravity. In other words, things, whether they are molecules or large objects, have certain inherent ways of behaving in certain surroundings. No-one 'makes' them behave in this way; they just do.

So it is with a mixture of methane, hydrogen, ammonia and water. When assaulted by electric sparks, these substances re-arrange their atoms to form sugars and amino acids. Other basic chemicals treated in similar ways have been shown to form nucleic acid components (purines and pyrimidines). It's the laws of physics, or in this case, chemistry.

Nevertheless, you may think that it was quite a convenient coincidence that the chemical compounds needed to build proteins and DNA just happened to be those that were produced by lightning strikes and cosmic ray bombardment of Earth's early atmosphere. It sounds a bit like finding some random pieces of rock which just happen to fit together to form a beautiful vase.

Proteins and DNA, and other essential chemicals needed for life to form, didn't exist at this time. The laws of physics and of chemistry dictated that amino acids, and purines and pyrimidines, could be formed spontaneously as a result of natural processes on Earth, just like they did in Miller and Urey's experiment described earlier.

It just so happened that these compounds were able to join together to form protein and DNA-type molecules. If they couldn't, then life would not have developed.

Using the vase analogy, we can imagine searching amongst random rock chippings in a quarry and eventually collecting some that could be formed into a vase. It may be unlikely but with luck and enough time it's certainly possible.

So was the formation of proteins and DNA just a lucky chance? Of course it was, but it only had to happen once, and it took about 1 billion years. It's the same with the random pieces of rock. Most won't form a vase but if you search long enough you may just find some that do.

We can make a small but interesting diversion here. It had always been thought that the chemicals in living creatures were fundamentally different from those in the inanimate world – rocks, minerals, air etc.

There was a supposed *vital force* that made living things alive and different from non-living things. This was the origin of the terms *organic (living)* and *inorganic (non-living)*. Note that this has nothing to do with so-called 'organic' food, which is a dubious and ill-defined term when used in this context.

This all changed in 1828 when a German chemist, Friedrich Wöhler (Fig. 26; 1800-1882), was doing some experiments with an inorganic substance called ammonium isocyanate. Upon heating, the ammonium isocyanate changed into urea, an organic compound. This was a revolutionary discovery – the first ever synthesis of an organic compound from an inorganic one, and it marked the beginning of the end of the vital force concept. From now on, it was clear that there was no fundamental difference between organic and inorganic compounds as far as any invisible vital force was concerned.

Fig. 26 Friedrich Wöhler in the 1850s (from Wikimedia Commons)

Going back to the origin of life, it's encouraging that the basic building blocks could have been produced by natural processes on the young Earth. However, the fundamental property of life is that it can reproduce itself. If it couldn't do that it wouldn't survive.

To start with then, what we need is a molecule that can produce a copy of itself. Once we've achieved that, the rest could follow.

Various theories have been put forward as to how this might have happened. Here's one. Think of some jigsaw pieces floating around in a lake or in the sea. Shaken about, some of them might interlock with each other to form a short chain. To help with the mental image, assume that all these pieces are grey. This short grey chain then encounters some more pieces, which are black (Fig. 27).

The black pieces might then interlock with the grey pieces to form a second chain, so that we how have a grey chain and a black chain linked together. If the joining parts of the black chain match those of the grey chain, and they come into contact with more grey pieces, then a second identical grey chain could be formed.

So what has happened here is that the grey chain has reproduced itself *with its constituent pieces in the same order as in the original chain.* This is basically how DNA works.

We know that the building blocks of DNA (the jigsaw pieces in Fig. 27) can be produced by natural processes, so this sort of scenario is theoretically possible.

Our conclusion therefore is that it is possible for the building blocks of life to form on their own under suitable conditions.

Fig. 27 Possible mechanism for self-replicating molecules
LEFT HAND GREY CHAIN forms by random collisions in a mixture
CENTRE BLACK CHAIN interlocks with left hand grey chain
RIGHT HAND GREY CHAIN forms in the presence of more grey pieces

If they were to encounter some fatty or oily materials, then perhaps a wall or membrane could form around them. This would then certainly resemble a simple cell which would have the ability to make copies of itself.

The processes described by Charles Darwin, and known as natural selection due to the survival of the fittest then took over, and evolution (which actually just means *change with time*) eventually resulted in the variety of living creatures seen today.

There are lots of *possibles* and *perhapses* here and the entire process may have been very unlikely. However, this doesn't matter – all we're concerned with is whether there is a plausible mechanism. How unlikely isn't important at the moment since it only had to happen once (see Question 7 for more on this).

Is it possible that life could have come into being on its own?
Yes, it is possible.

2 THE CREATOR OPTION

Genesis 1:1 In the beginning God created the heaven and the earth.

Genesis 1:11 And God said Let the earth bring forth grass, the herb yielding seed:

Genesis 2:2 And on the seventh day God ended his work which he had made;

The Book of Genesis at the beginning of the Old Testament was probably written around 1300 BCE, and taken from other writings and stories handed down through the generations. It states that God created heaven and Earth, and that he did it in 6 days, resting on the seventh.

Scientific advances during the 19th century led many people to abandon the literal interpretation of the Bible and to accept the view that the Earth and Universe were very old. By the early 20th century, Darwin's ideas concerning evolution were also becoming widely accepted.

In 1920s America, a Christian fundamentalist movement began which was opposed to the idea of evolution and succeeded in getting its teaching banned in public schools. The famous Scopes Monkey trial of 1925, later made into the film *Inherit the Wind (1960)* resulted in a school teacher being fined for teaching evolution to his class.

The term *Creationism* was coined as an alternative model to that proposed by evolution. In the 1960s, this became *Scientific Creationism* although still proposing a literal interpretation of the Book of Genesis. In the 1990s, this then became known as *Intelligent Design*, to strip it of any biblical references. It holds that intelligent intervention was necessary for the creation of life and of the Universe.

Today, a very large number of people of different religions still interpret the writings of the Bible literally, and therefore believe that the entire process of creation did in fact take just 6 days.

That's what it says in the book. But it's an old book. A very old book. You have to ask yourself, how many 3,000 year old ideas are still true today? Just because people thought something was true 3,000 years ago doesn't mean that you have to believe it now.

Would you want your doctor to be prescribing medicines from a 3,000 year old medical manual? *"Sorry, antibiotics haven't been discovered yet but this arsenic and locust potion may help."* Is the Earth flat? Do the Sun, Moon, planets and stars all revolve around the Earth? If you're buried with all your earthly possessions, can you use them in an after-life? Is everything made of Earth, Air, Fire and Water?

Creationism and Intelligent Design are fundamentally the same thing under different names. The belief is that life and the Universe were the work of a Creator or designer. Some adherents believe in a literal interpretation of the Bible and hold that the Universe is only about 10,000 years old. Others, while still believing in a Creator or designer God, do accept the scientific evidence for a very old Universe.

We'll come back to the age of the Earth and Universe in Question 4 but for now we're concerned with the God option for the creation of life.

We've seen that scientific evidence does exist to allow for the possibility that life arose on its own, this being a slow and gradual formation from simple self-reproducing compounds to simple cells and eventually to more complex organisms. To repeat, no-one is suggesting that finished complex living things suddenly appeared on their own.

As we shall see in Question 15, approximately 60% of the World's population believe in a supreme being, usually but not always called God, responsible for the creation of life on Earth. Since the remaining 40% do not believe in God, they presumably believe that life, somehow, arose without divine intervention.

It's possible that life arose on its own. It probably took about 1 billion years to get started and it may have been very unlikely but it only had to happen once. In that case, why complicate the explanation by saying that God, about whom we know nothing, did it by another unknown process?

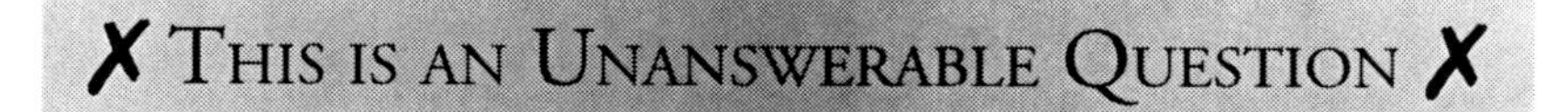
✗ This is an Unanswerable Question ✗

BEST GUESS ANSWER

It seems most likely that life arose on Earth after the basic building blocks (amino acids, sugars, and nucleic acid components), were produced as a result of lightning strikes and cosmic ray bombardment on the primitive Earth. These components came together by random collisions in mixtures, probably in the sea, and formed simple compounds capable of self-replication, eventually forming single cells. The processes of natural selection took over, eventually resulting in the variety of livings things seen today.

Question 4: How old are the Earth and the Universe?

We have two options.

1 **The Biblical option – thousands of years**
2 **The scientific option – billions of years**

The BIBLICAL option versus the SCIENTIFIC option

Creationism and Intelligent Design are fundamentally the same thing, the latter name having been coined to make the concept more acceptable. While all proponents of these ideas believe in a Universe created by God, they differ in other aspects. Some maintain that the Universe is as old as estimated by astronomers whilst others say that the biblical account is literally true, and that creation did take just 6 days, that God created each kind of life individually, and finally that it is no more than 10,000 years old.

If this is the case, then how is it that we can see stars and galaxies that are millions and billions of light years away? Imagine we're looking at a galaxy that is 50 million light years away. What this means is that the light from that galaxy takes 50 million years to reach us. If we can see it, then it, and the Universe, must be at least 50 million years old. The furthest galaxy currently known is over 13 billion light years away, so the Universe must be at least 13 billion years old. How do Creationists get round this problem?

It's called *in-transit creation*, and states that God created this light in transit about 10,000 light years away. This is then consistent with the biblical age of the Universe and also explains the apparent much older age measured by astronomers.

It also means that events seen and measured by astronomers, such as supernovae, other galaxies, black holes, quasars etc., don't really exist since the light (and radio waves and X-rays) by which we see and detect them were placed in space by God 10,000 years ago to give us the impression of a much older and more complex Universe.

Fig. 28 Andromeda spiral galaxy. © NASA

Fig. 28 is a photograph of the Andromeda spiral galaxy. It is 2.5 million light years away and on a clear night is just about visible to the naked eye as a faint smudge. Do you believe that it doesn't really exist and that its light was placed in space by God to fool you?

Why? What would be the point? If these objects really don't exist then why invent them? It makes no sense.

The case for a young Earth arises from a literal acceptance of the Bible, in particular, those portions of the Old Testament which give an unbroken male lineage from Adam to Solomon complete with the ages of the individuals.

Further analysis of events detailed in the Old Testament has resulted in a chronology from the Creation up to the birth of Jesus.

Several such chronologies have been published, and one of the best known is that of James Ussher, a 17th century Irish Archbishop. He provided an exact date for the Creation - **23 October 4004 BCE.** This value is so different from the scientific one of about **13.7 billion years** for the Universe, and about **4.5 billion years** for the Earth, that it should be easy to prove one way or the other. Here are a few arguments for a very old Earth and Universe.

Dating of rocks by radioactive decay
Ice core and tree ring data
Observance of galaxies billions of light years away

These arguments seem irrefutable. Radio-dating is a well-established scientific dating method, counting ice core layers and tree rings just needs time and care to make the counts, and being able to see galaxies billions of light years away means that their light must have started its journey many billions of years ago to get here for us to see it now.

However, all these seemingly solid arguments can easily be put aside. The young Earth creationists' response centres on two rebuttals.

i) The rate of radioactive decay we measure today may not be the same as it was in the past, thus giving false results. Since no-one was around to measure these parameters millions of years ago, it's not possible to say with absolute certainty that these rates were the same then as now. If they have in fact changed, then the age of the Earth could be 10,000 years rather than 4.5 billion years.

This is a clever argument, and to answer this we first need to describe in some detail the process of radioactive decay and the concept of *half lives.*

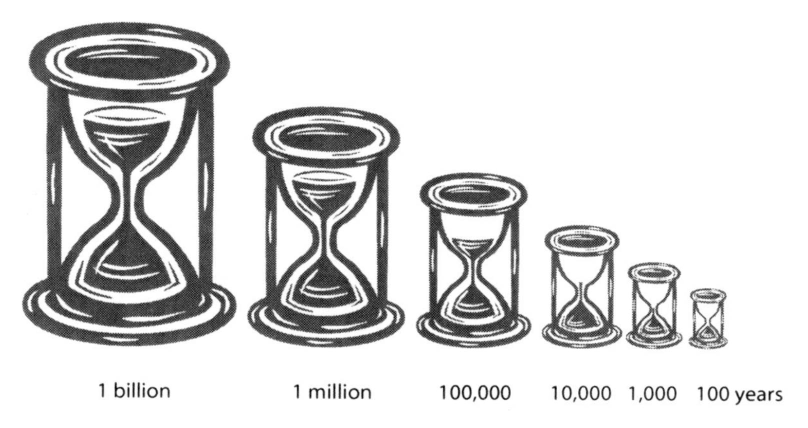

Fig. 29 Sand timers to illustrate the principle of radio-dating

Fig. 29 shows 6 sand timers. Each one runs for a different time as indicated. The big one on the left takes 1 billion years for all of the sand to fall to the bottom. The next one takes 1 million years, and then 100,000 years, 10,000 years, 1,000 years, and finally, the small one on the right takes 100 years.

Now imagine that you come across these timers standing in a row as in the diagram. You'll see that the 100 hour and the 1,000 hour ones are finished, and that the 10,000 hour one is just about to finish. The

100,000 hour one and the 1 million hour one both still have lots of sand left, while the 1 billion hour one has barely changed. What's your conclusion about how long ago they were set up?

It's pretty obvious that the answer is just under 10,000 years. The shorter ones have finished, and the longer ones haven't.

This example is meant to illustrate how the radioactive decay of elements can be used to date the Earth. First though, we need a brief description of radioactive decay.

Elements are made of atoms, and each atom consists of electrons spinning round a nucleus made of protons and neutrons.

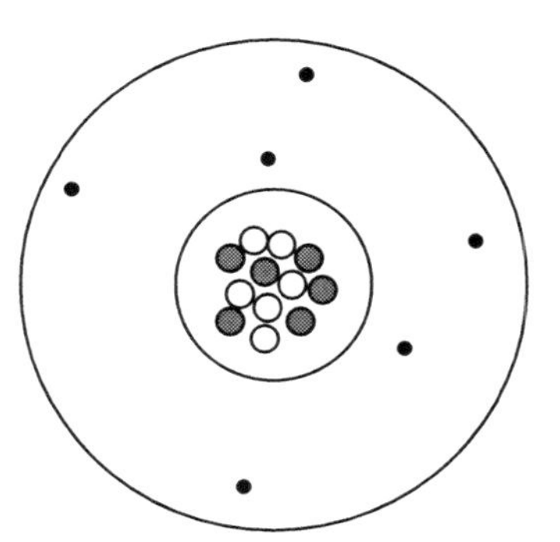

Fig. 30 Diagram of a carbon atom

Fig. 30 is a simplified drawing of a carbon atom. There is a central nucleus that contains 6 protons and 6 neutrons and spinning round the nucleus are 6 electrons.

Many atoms exist in multiple forms known as *isotopes*. These differ from each other in the numbers of neutrons in the nucleus. Carbon 14, for example, has 8 neutrons rather than 6 (8 neutrons plus 6 protons = 14 particles in total, hence carbon 14).

Some isotopes can be unstable, and over time will lose some of their electrons, protons or neutrons. This then results in the original atom, say uranium for example, becoming changed into another one (which in the case of uranium would be thorium). This process is known as *radioactivity* and the change from one element into another is known as *radioactive decay*.

One more thing needs to be explained and this is the *half-life*. This term means the time taken for half of a radioactive substance to decay into another substance. Half lives vary a lot. For example, the half life of uranium 238 is about 4 billion years, while that of nobelium 248 is 2 millionths of a second (the half-life is independent of how much material was there to start with).

Radioactive decay is very useful for geologists since it gives them a whole set of sand timers with different running times. They would have been set up when the Earth became a solid planet and the rocks had stabilised from their original molten state.

If half of something decays (disappears) in 1 million years, for example, then after 2 million years another half will have gone and only one quarter will be left (a half of a half), and so on. After 10 half lives (10 million years) only one thousandth (actually 1/1024) of the original substance will be left. Eventually a substance will become undetectable since there will only such a minute amount left that it can't be measured.

Here is a table of some radioactive isotopes (known as *nuclides* by chemists) with their half-lives. None of them are currently produced either by natural reactions or by cosmic rays – any found are therefore the remnants of what was originally present when the Earth was formed.

Isotope	Half life (years)	Found on Earth?
Vanadium 50	6,000,000,000,000,000	yes
Neodymium 144	2,400,000,000,000,000	yes
Hafnium 174	2,000,000,000,000,000	yes
Platinum 192	1,000,000,000,000,000	yes
Indium 115	600,000,000,000,000	yes
Gadolinium 152	110,000,000,000,000	yes
Tellurium 123	12,000,000,000,000	yes
Platinum 190	690,000,000,000	yes
Lanthanum 138	112,000,000,000	yes
Samarium 147	106,000,000,000	yes
Rubidium 87	49,000,000,000	yes
Lutetium 176	35,000,000,000	yes
Thorium 232	14,000,000,000	yes
Uranium 238	4,500,000,000	yes
Potassium 40	1,250,000,000	yes
Uranium 235	704,000,000	yes
Samarium 146	103,000,000	no
Plutonium 244	82,000,000	minute traces
Niobium 92	35,000,000	no
Curium 247	16,000,000	no
Lead 205	15,000,000	no
Hafnium 182	9,000,000	no
Palladium 107	7,000,000	no
Caesium 135	3,000,000	no
Technetium 97	3,000,000	no
Gadolinium 150	2,000,000	no
Zirconium 93	2,000,000`	no
Technetium 98	2,000,000	no
Dysprosium 154	1,000,000	no

You will notice that the isotopes with half lives greater than about 100 million years can be detected on Earth whereas those with half lives less than this figure cannot be detected. (Plutonium 244, with a half life of 82 million years has been detected in absolutely minute quantities because a group of geologists made a supreme effort to see if they could find it. If someone tried just as hard to try and find samarium 146 they too might be successful).

A reasonable question might be – if the isotopes in the bottom part of the table are not detectable on Earth, then why do they have names and how do we know what their half lives are? All of these substances can be produced in nuclear reactors and other atom smashing machines and can therefore be studied.

They would have existed on Earth at one time but, as explained above, have all decayed into other products (just like the sand in the last three timers has all run into the bottom chamber).

We can now return to our original question about the age of the Earth. The scientific age is about 4.5 billion years while the new Earth Creationist age is about 10,000 years. There is a break in the half life table somewhere in the region of 80 to 100 million years. Isotopes with half lives longer than this can be detected on Earth whereas isotopes with half lives shorter than this cannot be detected.

If the Earth was 10,000 years old, then even the last isotope in the table, dysprosium 154 with a half life of 1 million years, would be abundant since it would not even have been around for one half life. A 10,000 year old Earth would contain every isotope in the table, and a great many more with shorter half lives as well.

On the other hand, a 4.5 billion year old Earth would be expected to have depleted its original stock of isotopes with half lives less than about 100 million years since this represents about 45 half lives (45 x 100 million years = 4.5 billion years). After this time, there would be only about one 30 trillionth of the original substance left, and this would be undetectable.

The evidence is therefore completely compatible with an Earth age of around 4.5 billion years and completely incompatible with an Earth age of 10,000 years. However, what about the notion that the decay rates have changed or that they are wrong or that the calculations used by physicists and geologists are flawed.

Let's examine this idea. A well-known method for dating trees is to count the annual rings (Fig. 31). Ice cores can also be dated by counting the ice layers.

Fig. 31 Tree rings

Radio-dating techniques and dendrochronological (tree ring counting) techniques and ice core counting techniques all give the same answers up to several thousands of years. So here we have two completely different methods based on completely different principles that give the same results.

A good analogy would be a comparison between two different types of scales, as shown in Fig. 32.

Fig. 32 Two different types of scales

The scale on the left is a traditional type of balance device based on the lever principle. When the weights in both pans are the same, the beam is horizontal as indicated by a pointer in the middle of the beam. The weights in the right hand pan are then equal to the weight of the item in the left hand pan.

The scale on the right depends on the depression of the base. The base is connected to the dial pointer by a series of springs and connectors; the greater the depression (weight) the more the dial moves.

The point is that these two weighing devices are based on completely different principles and are therefore extremely unlikely to suffer from the same errors. If an item weighs the same on both scales we can be confident that this is its true weight.

It's the same with the two different methods for measuring the ages of trees and ice cores by ring counting and by carbon dating. If both methods give the same result, we can also be confident that the results are correct.

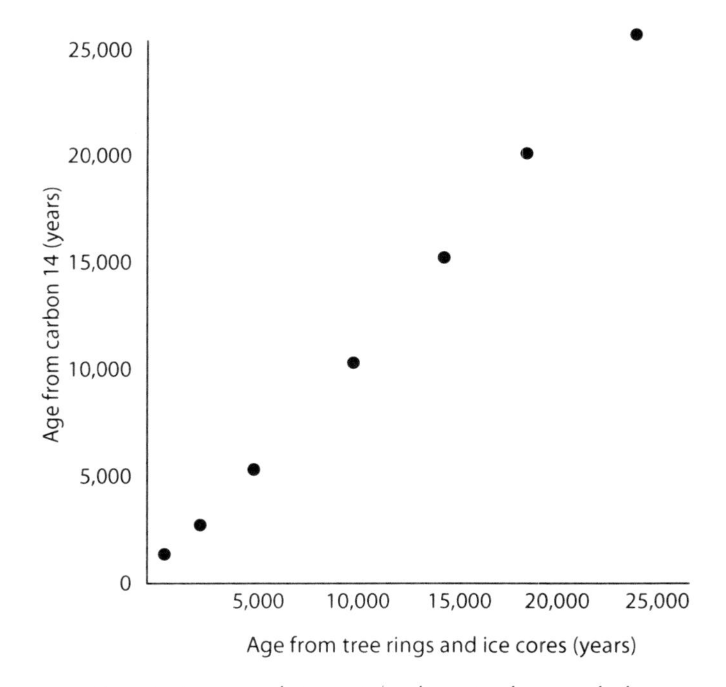

Fig. 33 Comparison between C14 and tree ring dating methods (data compiled from IntCal04 figures)

Fig. 33 is a calibration graph where tree rings and ice cores have been counted to determine their ages back to 25,000 years. The same tree samples and ice cores were also dated by the carbon 14 method. The correlation is excellent leaving no doubt whatsoever that the carbon 14 dating method is accurate to at least 25,000 years.

As was discussed earlier in this chapter, we know that isotopes with half lives of less than about 100 million years are no longer detectable on Earth because they have been through so many half lives that the amount left is too small to measure. How could the absence of, say, niobium 92 with a half life of 35 million years, be compatible with a 10,000 year old Earth?

For a radioactive isotope to become undetectable, it would have to have been through about 40 or 50 half lives.

In 10,000 years, the amount of niobium 92 would hardly have changed at all, and there should be lots of it about. For it to have become undetectable in just 10,000 years, its half life would have to be about 200 years (because 50 half lives in 10,000 years implies a half life of 200 years) rather than 35 million years. That's a staggering difference from what scientists have observed. It's a difference of 175,000 times. Similar calculations apply to all the other radioactive isotopes listed in the table that have become undetectable.

It's just inconceivable that all the half lives that have been measured are wrong by such large factors. Bear in mind that the half life of carbon 14 is known to be correct since it's been independently calibrated against known ages of trees and ice cores. Why should this be right and so many others be wrong?

We can also look at some rock samples (from Norway) that have been dated by different people using different isotopes, as shown below. They have all given pretty much the same results. All the samples were from the same rock, and they were dated by using different isotopes, each with different half lives and different decay mechanisms.

Again, it's inconceivable that they have all given the same wrong results.

Dating Method	Calculated Age
Argon 40/Argon39	588 million years
Potassium/Argon	575 million years
Rubidium/Strontium	578 million years
Lead/Lead	573 million years
Thorium/Lead	580 million years

from an article by Joe Meert (2000); gondwanaresearch.com/radiomet.htm

Say, for example, you were timing a race and someone said that the time was wrong because your watch was running too fast or too slow.

So it's done again but this time you have 5 people each with their own timepieces. One is quartz, one is mechanical, one is kinetic, one is radio-controlled, and one is an atomic clock. All the five results are very similar. What do you conclude?

Are you going to say that all these timepieces, with different mechanisms, all have the same error so that they all give the same but wrong time? Of course not.

It's pretty clear therefore that radioactive decay is a valid technique for dating rocks and other materials. But what if you still don't believe? What if you still think that the decay rates were different millions and billions of years ago and all the results are wrong. After all, no-one was around then to measure the rates.

Actually, that's not quite true. We do have access to a remarkable time machine that can help us here.

When you look at the Moon you're seeing it as it was about one and a half seconds ago since that's how long it takes for the moonlight to travel the 240,000 miles from the Moon to the Earth.

So you're actually looking back in time by one and a half seconds. With the Sun, it's just over 8 minutes. And with the Andromeda galaxy, it's 2.5 million years.

Because astronomical objects are so far away, the light by which we see them takes so long to get here that we are looking at these objects as they were when the light left. For Andromeda, that's 2.5 million years ago. We don't know, and have no way of knowing, whether Andromeda is even there *now*.

A star can explode for various reasons, and when it does so it can form what is called a *supernova*. This is an extremely bright object, and many were observed by ancient astronomers hundreds of years before the telescope was invented.

Complex chemical reactions take place inside a supernova and these result in the formation of many radioactive isotopes. These isotopes can be detected and analysed by changes in the light emitted by the supernova.

One such supernova that has been studied in great detail is known as SN1987A. It's in a galaxy close to the Milky Way and is 169,000 light years away. That means that the light that we see it by is 169,000 years old. Spectroscopic analysis of this light shows that the decay rates of the radioactive isotopes produced by the supernova 169,000 years ago are the same as those measured here on Earth now. Similar results have been obtained from other supernovae much further away.

This finally nails the question of decay rates changing over time. They don't.

This has been quite a detailed section since it's important to understand the science behind radioactive decay to appreciate the evidence in its support.

The Earth is around 4.5 billion years old. Anyone who rejects the evidence in favour of this either doesn't understand the principles of radioactive decay or is deliberately ignoring the evidence.

The concept that the Universe had a finite age, rather than having existed forever, only arose in the 1920s. The American astronomer Edwin Hubble discovered that galaxies were large accumulations of stars outside and far away from our own Milky Way. He discovered that all the galaxies were speeding away from us and from each other, and the further they were away, the faster they were receding.

Calculating backwards, it's possible to arrive at a time when all of the galaxies, and all of the matter in the Universe, existed as a single point that then exploded in the Big Bang that created the expanding Universe we see today. This time is about 13.7 billion years ago. It's consistent with the estimated age of the oldest stars, about 13 billion years.

Mention should also be made of another calculation that gives a similar result. As was explained in Part One A, carbon, which is an essential element for our type of life, is produced by reactions inside stars, and is then distributed into space by super nova explosions as some stars die. This is the only source of carbon, and without it, we wouldn't be here.

From what is known about the reactions that occur inside stars, it has been calculated that this process of forming and distributing carbon (and other heavier elements) throughout the Universe takes around 10 billion years.

This is not an exact calculation but even the approximation is within the ball park of the scientifically accepted Universe age of 13.7 billion years.

To consider that it could have been done within one millionth of that time period – 10,000 years – goes against everything that is known about the reactions that occur within stars.

The second rebuttal, although seemingly preposterous, is just impossible to refute.

ii) God made it seem as if the Earth and Universe were very old by putting in place the right amounts of radioactive compounds, tree rings, ice cores, and the light of galaxies 'in transit' so that scientific measurements would be fooled into giving incorrect answers indicating a very old Earth and a very old Universe.

Why? What (on Earth) would be the point? We can't refute this but we can employ Ockham's Razor and ignore it.

✓ This is an Answerable Question ✓

ANSWER

Dendrochronology shows that some trees are several thousands of years old and these ages correlate exactly with those obtained by radio dating of the same tree samples with carbon 14. Radioactive isotopes with half lives of about 100 million years or less are undetectable on Earth. They would have to have been through about 50 half lives for this to happen, which equates to a time period of up to 5 billion years. For such isotopes to have become undetectable in 10,000 years, their half lives would have to be about 200 years which is a staggering error in measurement or change in decay rate.

Also, and very convincingly, different isotopes all give the same dates for the ages of old rocks. Finally, half lives measured from the light of distant supernovae confirm that the rates measured on Earth now are the same as they were hundreds of thousands and millions of years ago.

Distant galaxies have been measured as being billions of light years away, meaning that the Universe must be billions of years old. The most distant object known, which was detected by a NASA satellite on 23 April 2009, is known as GRB 090423 (gamma ray burst 2009 April 23). The light from this stellar explosion took 13 billion years to reach the Earth and therefore started its journey when the star exploded 13 billion years ago. That means that the Universe must be at least 13 billion years old. Calculating back from the known present expansion of the Universe actually gives the date of formation of the Universe as about 13.7 billion years ago.

James Ussher's and other people's chronologies are based on a literal interpretation of Biblical references and an extrapolation of timelines based on quoted ages. For example, in *Genesis 5:26* it states that *all the days of Methuselah were 969 years, and he died.*

The oldest person who ever lived with proper documents to prove it was a French lady called Jeanne Calment, who was born in 1875 and died in 1997 aged 122 years and 164 days. The notion that a human being could live for nearly 1,000 years, even in modern times, is unimaginable.

The explanation, presumably, is that either the biblical 'year' isn't quite the same as our modern year or that the biblical narratives are incorrect. Whatever the actual reason, the chronologies are invalid.

The Earth is about 4.5 billion years old, and the Universe is about 13.7 billion years old.

Question 5: Do UFOs exist?

A Texan newspaper published a story in 1878 about a local farmer who described seeing a circular flying object that was about the size of a saucer, flying at a 'wonderful speed'. This is believed to be the origin of the term 'flying saucer'. The United States Air Force coined the term Unidentified Flying Object, UFO, in 1952 to describe any such flying objects that remained unidentified after expert scrutiny.

There have been many thousands of reported UFO sightings and they continue to be reported both by people on the ground and also by airline pilots during flights.

So what are they? Most are probably due to atmospheric or climactic conditions such as unusual cloud formations, unusual lightning patters, astronomical events, weather patterns, weather and military balloons or other aerial devices, or merely mirages or imaginary sightings. There remains however a large body of people who insist that at least some of the sightings are of real aerial devices with the suggestion that they are of extra-terrestrial origin.

Fig. 34 shows a typical photograph of a UFO sighting.

Numerous such photographs have been published. The difficulty is that they are so easy to fake that most people attach very little credence to them.

Fig. 34 UFO sighting in New Mexico, USA, 1967
© The Paranormal Borderline

In 2010, the National Archives released a large number of previously restricted files on UFO sightings. Included in this archive were many drawings that had been sent in by members of the public. Fig. 35 shows a small selection.

UFO 1 1954; 30 feet wide with flashing blue lights.
UFO 2 1993; aliens seen through aircraft window
UFO 3 1997; side (above) and front views; covered in lights
UFO 4 undated; no other details
UFO 5 1980; no other details
UFO 6 1994; 40 feet long; 20 feet wide; hovered 10 -15 feet above ground for 40 minutes; no lights; no sound
UFO 7 1995; observed 20 feet away sucking up water
UFO 8 undated; 45 feet wide

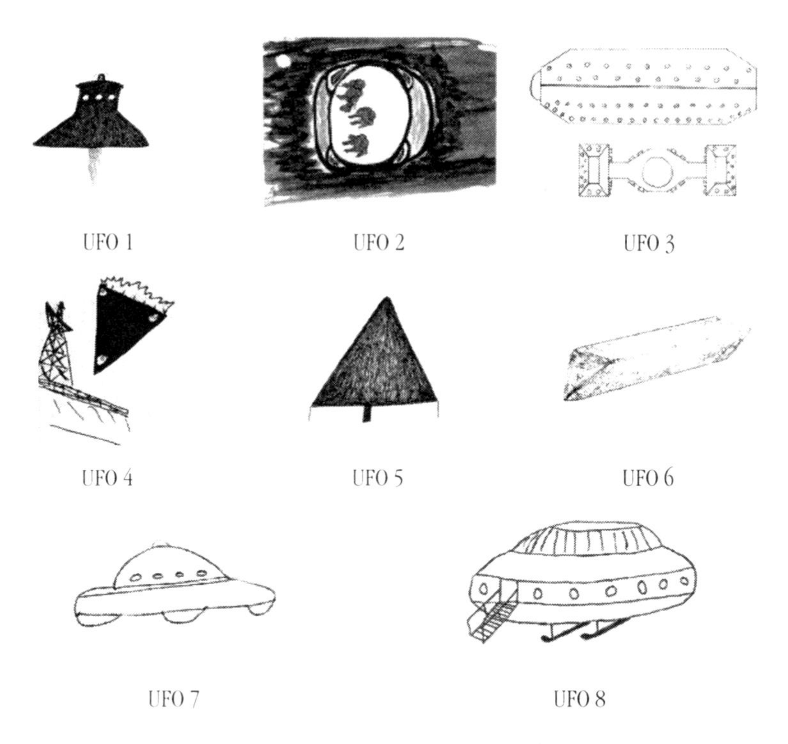

Fig. 35 UFO drawings submitted by members of the public.
By permission of the National Archives 2010

Even this small selection encompasses a wide variety of shapes and designs. The free-floating aliens (UFO 2) must have been a remarkable sight for this airline passenger. The Toblerone-shaped object (UFO 6) is the length of a London bus and nearly three times as wide. What a pity no-one thought to take some photographs as it hovered 10 feet above the ground for 40 minutes. Note the helpful stairway and sleds on UFO 8. The triangular shapes of UFO 4 and UFO 5 are reminiscent of the American Stealth bomber (Fig. 36).

Fig. 36 Drawing of U S Air Force Stealth bomber copied from a photograph of the actual airplane

The American authorities would have encouraged members of the public to think that they had seen a UFO rather than a prototype secret airplane.

A similar selection of drawings was released by the New Zealand Air Force in 2010.

Perhaps the best known UFO event was the **Roswell Incident**.
This occurred in the summer of 1947 near the town of Roswell in New Mexico, USA, when the Roswell Army Air Field (RAAF) announced that it had recovered debris from a crashed 'flying saucer' (Fig. 37). Subsequent reports changed this, stating that the debris was from a high flying weather balloon.

UFO enthusiasts seized upon this story which soon became embellished with reports of alien corpses being secreted away by the US military. There have been numerous books, TV programmes and films about this event, and although many people accept the balloon interpretation, there are also others who are convinced that the US military have concocted a cover story to hide the truth about this event.

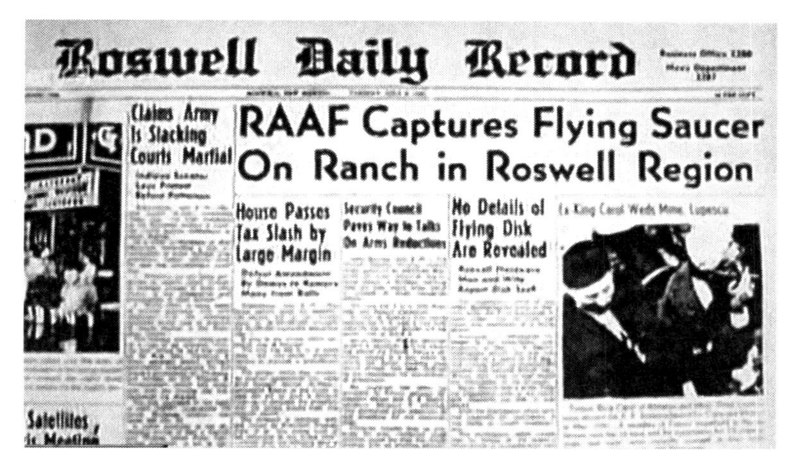

Roswell Daily Record

RAAF Captures Flying Saucer On Ranch in Roswell Region

Claims Army Is Stacking Courts Martial

House Passes Tax Slash by Large Margin

Security Council Paves Way to Talks On Arms Reductions

No Details of Flying Disk Are Revealed

Fig. 37 Newspaper Headline, 8 July 1947
© Roswell Daily Record

It is certain that something fell out of the sky and it is also certain that the US military were conducting secret work with experimental flying craft with a view to spying on the Soviet Union. It would suit them to support the notion of UFOs to deflect attention from their clandestine activities, and the crash of a secret military device does seem a much more likely explanation that the crash of an alien spaceship. Even so, the story persists.

Experimental **circular aircraft** could have been a source of many UFO sightings in the 1940s and 1950s.

Fig. 38 shows a painting of a prototype German flying disc, sometimes known as the V7 or the Bellonzo-Schriever-Miethe Disc. It reportedly flew in Prague, Czechoslovakia, in 1945 but was never developed further (or may not even have existed!) There is much discussion about these craft on the internet.

Fig. 38 Bellonzo-Shriever-Miethe flying disc.
Original painting © Jim Nichols 1990

Fig. 39 shows an Avro Canada VZ-9 Avrocar, another experimental and secret circular aircraft built for the US military in the late 1950s. It never reached its design specifications, and after various trials and structural modifications, it was retired in 1961. Only two craft were ever built.

Fig. 39 Avro Canada VZ-9 Avrocar (1958 – 1961)
(from Wikimedia Commons)

Anyone observing these secret airplanes in flight could be forgiven for thinking that they had seen a flying saucer.

Unusual **cloud formations** are probably a very common source of reported sightings. *Lenticular clouds* are lens shaped cloud formations that can look strikingly like what many people would call a flying saucer (Fig. 40).

Fig. 40 Lenticular clouds
(Wikimedia Commons)

Photographs are easy to fake, drawings may or may not represent what was seen, and some sightings might look like strange flying craft but could be secret military prototype aircraft or atmospheric phenomena. But let's for a moment imagine that at least some of the sightings are in fact real alien spacecraft. Two questions arise.

Where are they from?

Spacecraft have visited every planet and many of the moons in our Solar System with the exception of Pluto, where the surface temperature is about -230°C, and found no evidence whatsoever of any life. A civilisation sufficiently advanced to be able to send space craft to Earth would surely have some traces of their existence on their home planet which would have been detected by our space crafts' cameras.

So we have to look further away. The nearest star to Earth with probable planets, although these are still to be confirmed, is Epsilon Eridani (data from *The Extrasolar Planets Encyclopedia – http://exoplanet.eu/*) which is 10 light years away. It seems therefore that any aliens would have had a journey of at least 60 trillion miles to get here. It sounds a long way to us but for them might just be the equivalent of a walk in the park. All we do know is that no radio transmissions indicating an intelligent source have been received from any cosmic source, and we would reasonably expect an alien race with sufficient technology to undertake such a journey to be broadcasting in search of other living creatures.

WHY IS THERE NO PHYSICAL EVIDENCE?

Although some pilotless space probes have been deliberately destroyed on arrival, such as the Galileo probe sent into Jupiter's dense atmosphere at over 10,000 mph after its 14 year mission was complete, most end up on the surface of their planet or moon, or at least in orbit around it. Two probes with no plans to land anywhere, Pioneer 10 and 11 which are now on their way out of the Solar System, carried plaques (Fig. 41) in the event that they are ever intercepted by an alien civilisation at some future time.

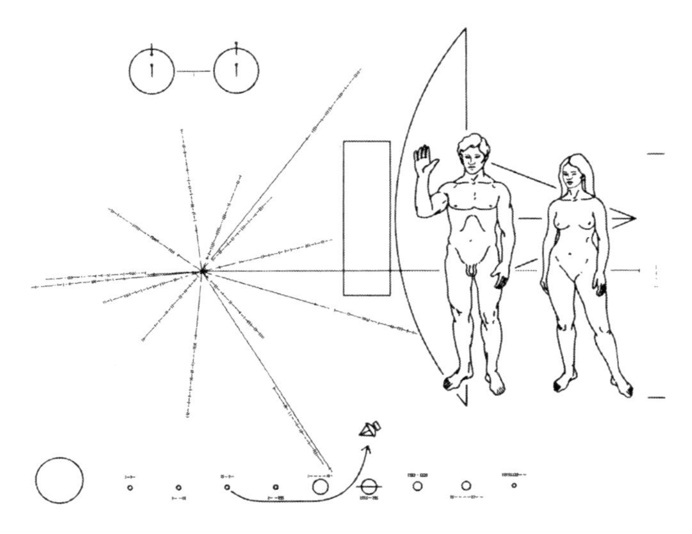

Fig. 41 Plaque attached to Pioneer 10 and Pioneer 11 spacecraft
© NASA

The message is intended to communicate the location of the human race, the appearance of an adult male and female of our species, and the approximate era when the probe was launched.

A line-drawing of a naked couple standing in front of the Pioneer probe is accompanied by an ingenious scheme for conveying information about the spacecraft's origins. For more information on the meaning of the symbols on the plaques see the Wikipedia article *Pioneer Plaque*.

It's inconceivable that visitors from another star system would come all the way here and not leave some indestructible evidence of their arrival, as did the first Apollo astronauts in 1969 (Fig. 42).

Fig. 42 Plaque on the landing gear of the Eagle spacecraft left on the Moon by the Apollo 11 astronauts in 1969. © NASA

So what are we to make of the numerous photographs, drawings and verbal reports that purport to show UFOs? In the absence of any credible supporting evidence, it's reasonable to assume that the observers were either genuinely misled in what they saw due to atmospheric or other interference, or that they did see what they reported but it was a local military or scientific device, or that they have deliberately fabricated their 'evidence' for their own purposes.

✓ This is an Answerable Question ✓

ANSWER

Reports of Unidentified Flying Objects have been made for well over 100 years. Every one is probably due to natural or man-made processes and logically explainable or due to deliberate falsification. There is no physical evidence for any extra-terrestrial visitations, and based on SETI research, there is no obvious origin for such visitors who at best would have to make a journey of 10 light years to reach Earth. It's highly likely that all UFO sightings are due to terrestrial objects or fabrications.

Question 6

Have alien astronauts visited Earth?

The idea that the Earth has been visited by alien creatures in the distant past who then influenced human development and technology is an intriguing one that was first developed in the late 19th century by various science fiction writers. Its best known proponent is the Swiss writer Erich von Daniken who published his book *Chariots of the Gods?* (originally in German) in 1968 (Fig. 43). According to his website (www.evdaniken.com), von Daniken has sold over 60 million books in over 30 different languages.

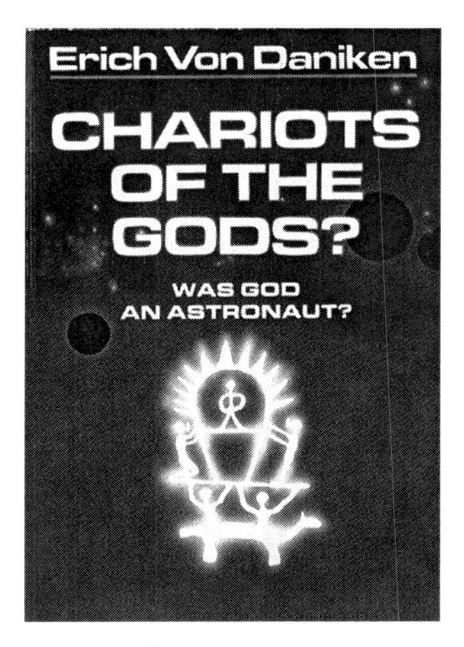

Fig. 43 Chariots of the Gods?
© Erich von Daniken

The Ancient Astronaut idea is therefore a popular concept and it has spawned many publications from many authors. The primary argument in favour of this thesis is that the Earth is littered with numerous artefacts that represent a higher level of technological expertise than existed at the time they were created. Supporting evidence is supposedly given in pre-historic drawings, some of which are said to resemble modern astronauts and space vehicles.

We can examine a few of the examples given in *Chariots of the Gods?*

Sacsayhuaman is a large walled complex near the city of Cusco in Peru (Fig. 44). It was built by a pre-Inca people in about 1100, and is remarkable for its walls which consist of numerous large stones that fit together so perfectly that in some cases it is not even possible to insert a sheet of paper between them.

Fig. 44 Part of the complex wall in Sacsayhuaman, Peru

Von Daniken refers to Sacsayhuaman in his book. He states "*Just look at the incredible accuracy of the jointing. How could primitive people handle these huge blocks?*"

Similar arguments are used for the Maoi statues on Easter Island, Stonehenge in the United Kingdom, and the pyramids of Egypt, to name but a few.

Look at the beautiful death mask of Tutankhamun (Fig. 45) created out of solid gold and coloured glass and semi-precious stones by ancient Egyptian craftsmen 3,400 years ago.

Fig. 45 Death mask of Tutankhamun circa 1330 BCE in the Cairo museum

This, and all the other artefacts recovered from ancient Egyptian tombs, were created by local craftsmen who had developed their skills over many years. Or do you think that an alien stood by to oversee the work?

These are indeed all marvellous structures but that doesn't mean that they needed extra-terrestrial assistance in their construction.

It's a big mistake, and an insult, to assume that ancient civilisations were stupid. What they lacked in modern tools and technology they made up for in an enormous labour force, no minimum wage or late penalty clauses, and an absence of health and safety legislation. They could employ thousands of workers on long shifts – if someone was injured or killed, it didn't matter. In the end, the job got done.

They had architects and craftsmen who learnt their trade by experimentation. Mistakes did happen though. If a doorway collapsed, it would be cleared and re-built in a slightly different way. If a window was created in the wrong place (Fig. 46) it would be blocked up.

Fig. 46 Builder's error showing a partial window blocked up (white dots) in Machu Pichu, Peru (*circa* 1450)

A well-known builder's error became manifest in 1178, five years after the construction of the Tower of Pisa had begun (Fig. 47). A flawed original design of a completely inadequate foundation meant that the 55 metre high tower, weighing about 15,000 tons, would never remain stable and it started to lean when construction reached the third floor.

Fig. 47 Leaning Tower of Pisa

So not all ancient monuments are perfect.

Pre-historic art features prominently in Ancient Astronaut theories and has been used as evidence of first-hand encounters with alien space travellers. Here's a typical example – a 10,000 year-old rock drawing found in Val Camonica, Italy – also mentioned in *Chariots of the Gods?* (Fig. 48).

The drawing shows two human figures with headgear that does indeed resemble a space helmet with protrusions that could be interpreted as antennae. But that's the problem – interpretation. All manner of masks and headgear were created by ancient civilisations in their homage to the gods, and this could just be one more.

Fig. 48 Rock carving, Val Camonica, Italy c 8000 BCE . United Nations World Heritage site

It is tempting to compare these rock drawings with photographs of floating astronauts (Fig. 49), and yes, the comparison is impressive (especially when the angles are adjusted to match!). However, that doesn't mean that it is true.

Fig. 49 American astronaut. ©NASA

Look at the two photographs in Fig. 50.

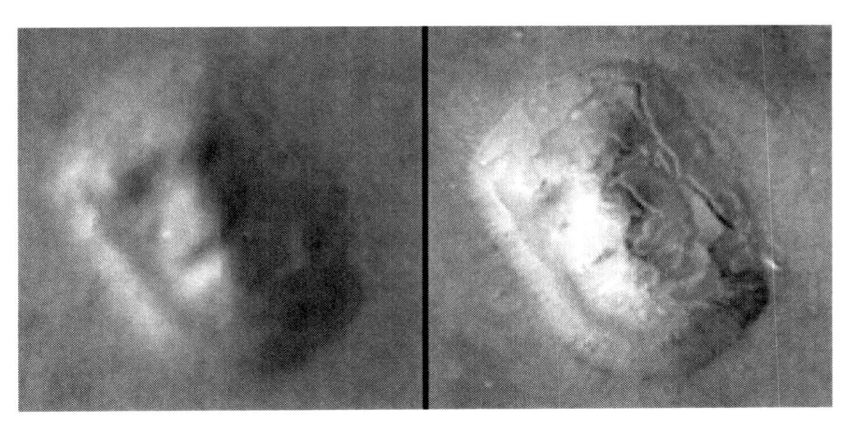

Fig. 50 'face' on Mars 1976 (left); 1998 (right)
© Malin Space Science Systems/NASA

In 1976, the Viking 1 Mars spacecraft was searching for possible landing sites for future missions when it photographed what has become known as the 'Face on Mars' (Fig. 50). The image did indeed resemble a human face and when released by NASA fuelled a great deal of speculation that it had been artificially created by a Martian civilisation. In view of the intense public interest in this phenomenon, NASA arranged for its Mars Orbiter Camera to take higher resolution images of the face in 1998.

It turned out that the 'face' was nothing more than a rock formation that, under suitable lighting conditions and low resolution imaging, had a passing resemblance to a human face.

There are quite a few other features on planetary mission photographs than have been put forward as evidence of extra-terrestrial life but all have turned out to be rock formations that, under certain lighting conditions, seem to resemble a human form.

Other artefacts that have been proposed as evidence of past extra-terrestrial visitations include **Ancient maps**. One such item has been prominent in this regard; it is known as the *Piri Reis* map (Fig. 51).

This map was drawn on a gazelle skin by Piri Reis, a Turkish admiral and map maker, in 1513. A fragment was discovered in the Topkapi Palace in Istanbul, Turkey, in 1929. Marginal notes on the map describe how it was complied from a variety of earlier source maps as well as from discoveries made on contemporary voyages by explorers and by ships blown off course.

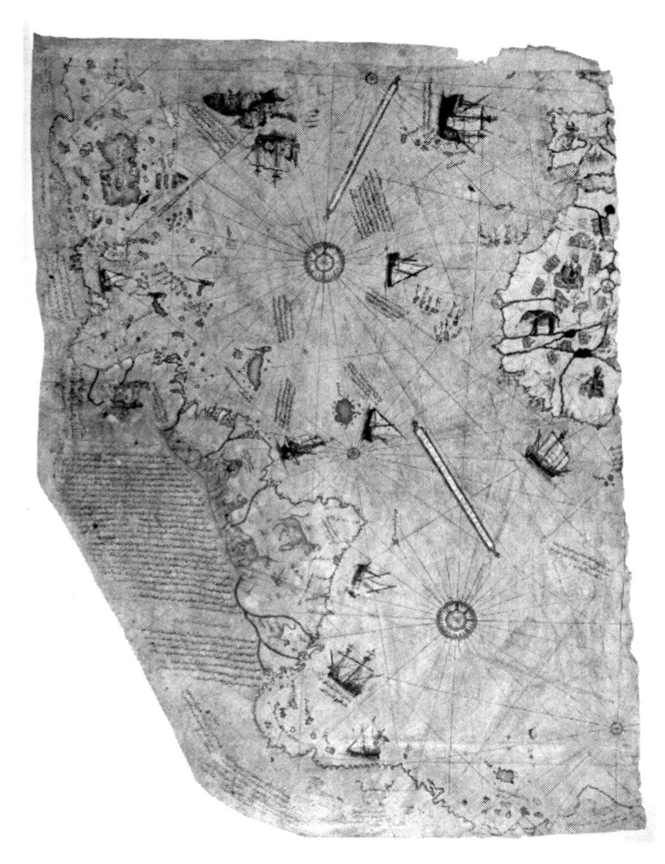

Fig. 51 Fragment of the Piri Reis map, drawn in 1513 © Topkapi Palace Library, Istanbul, Turkey

Some proponents of the Ancient Astronaut theory have suggested that the Piri Reis map contains information that could not have been known by humans in 1513, notably the existence of Antarctica. However, if we look at the sketch maps in Fig. 52, it is clear that although the upper part of South America is quite accurately mapped by Piri Reis, the coastline suddenly diverges eastwards at about the location of Rio de Janeiro.

To interpret this rogue coastline as representing the continent of Antarctica is fanciful. It's in the wrong place, is the wrong shape, and is shown as continuous with South America whereas in reality Antarctica is separated by the 600 mile wide Drake Passage

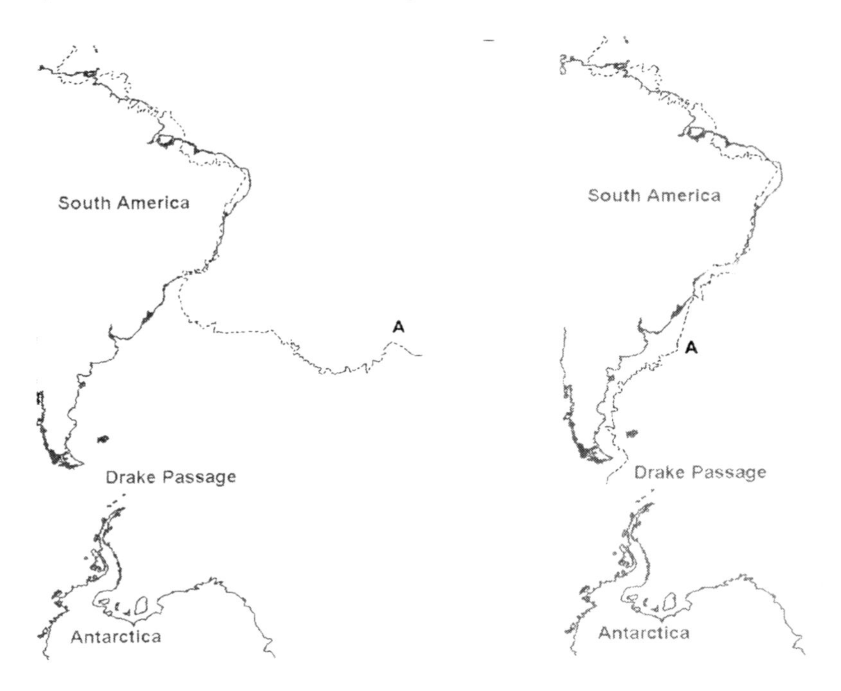

Fig. 52a (left) Modern sketch map of South America and Antarctica with part of the Piri Reis map super-imposed (dotted line A).

Fig. 52b (right) Showing the divergent coastline (A) bent downwards next to the actual coastline

So why was the coastline drawn that way? It is possible that the divergent coastline actually represents the rest of the South American coastline down to Cape Horn but bent just to fit onto the available space on the gazelle skin. Cartographers sometimes employ this device to make maps fit into available space. A more recent example of this is seen in the London Underground map from 1908 shown in Fig. 53.

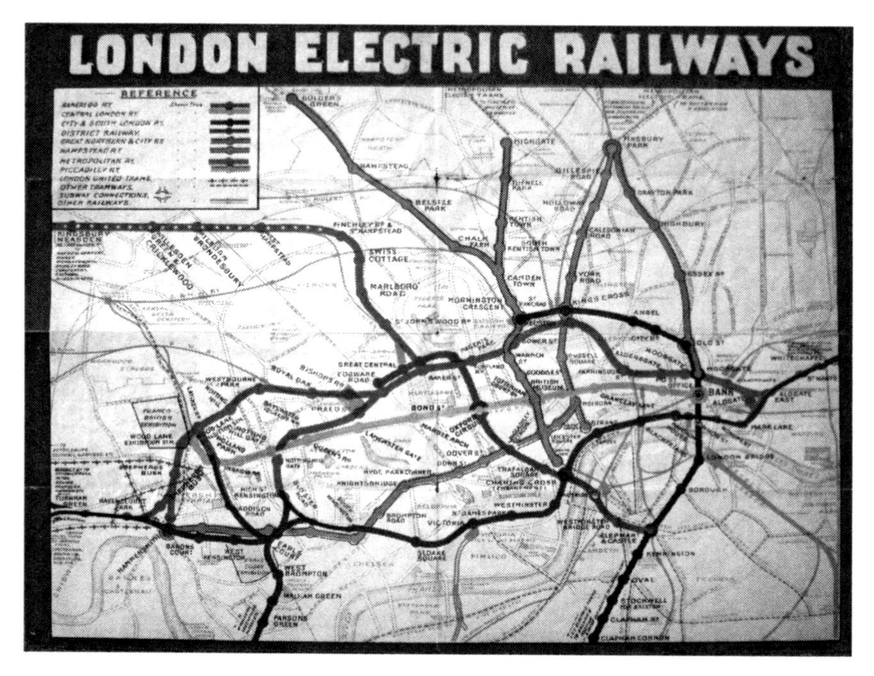

Fig. 53 London Underground map (1908).

These early London Underground maps showed the tube lines superimposed on a street plan and therefore gave a true representation of the geographical locations of the stations. This design lasted until 1933 when the map was radically re-drawn by Harry Beck in a diagrammatic style which was easier to read but no longer gave a true geographical representation of locations and distances.

The designer of the 1908 map faced a problem as to where he or she could locate the reference chart showing which colours represented which lines. It was eventually placed in the top left corner which meant that the western section of the Metropolitan Line, indicated by white dots in Fig. 53, had to be pushed downwards to create the necessary space. So even though the map purports to be geographically accurate, the cartographer had to adjust geography to fit in the reference chart. Perhaps Piri Riess had a similar problem of space – his gazelle skin just wasn't big enough.

It should also be mentioned that there had been a belief in a large southern continent since the Ptolemaic times of the first century, and depictions of large southern land masses were common in 16th and 17th century maps. Definitive evidence was eventually obtained by the first sighting of the continent in 1820. So even if Piri Riess's map had shown an Antarctic continent, which it doesn't, it would not have been something new.

Ancient statues of strange beings are another class of item cited by von Daniken and others as evidence of extra-terrestrial visitations, the idea being that these carvings were made in the image of the alien visitors. There are numerous such statues around the World differing in design, size, and date. A few examples are shown in Figs. 54, 55 and 56.

It's true that we have no knowledge as to where these ancient sculptors got their inspiration from but we shouldn't assume that they were devoid of imagination. Look at some of our own cultural artefacts, the Mount Rushmore Memorial in South Dakota, USA, and the statue of Eros in Piccadilly Circus, London for example (Figs. 57 and 58).

Fig. 54 Dogu statue, Jomon Period, Japan c1000 BCE. Wikimedia Commons

Fig. 55 Stone statue Kalasasaya Temple, Tiahuanaco. Bolivia c500 BCE

Fig. 56 Moai statues, Easter Island c 1500 (from Rivi @ Wikimedia Commons)

Fig. 57 Mount Rushmore memorial USA

Fig. 58 Eros statue, Piccadilly Circus, London ©BasPhoto-Fotolia.com

What might our descendants think of these giant heads and the flying archer should they be re-discovered in a few thousand years' time? We would surely laugh if we thought that they would believe them to be likenesses of 20th Century alien astronauts.

Finally, we can examine some **ancient Egyptian hieroglyphs** carved into a stone slab in the temple of Seti I in the ancient city of Abydos in about 1300 BCE. Fig. 59 shows a photograph of some of the hieroglyphs, and in particular, one that resembles a modern helicopter.

Fig. 59 Ancient Egyptian 'helicopter' (top left)

The image, which is widely circulated on the internet, does indeed look like a helicopter. You can see the rotor blade with the engine housing underneath, the tail fin, the body of the craft and the nose, and there is even an opening for entry. Now look at Fig. 60. This is an original *unretouched* photograph of the same area.

Fig. 60 Original unretouched photograph

You can see how the 'helicopter' (and some of the other glyphs) have been cleaned up in Fig. 59 by removing other touching and overlapping symbols. Scholars interpret the apparent helicopter as being the result of erosion of the original glyphs and later adjustments or re-writing over the original inscriptions. Parts of the older original symbols remained, and together this gives the illusion of a helicopter.

However, we don't really need a scholarly explanation. Even though the symbol may look something like a helicopter, do you really believe that the ancient Egyptians were flying around in these machines 3,000 years ago? Given their obsession with building temples and tombs, and their proficiency in carving, don't you think that such a wonderful machine would figure rather more prominently in their extant writings than on a single slab on a pharaoh's tomb?

There is no credible evidence for alien visitations to Earth in ancient (or modern) times.

✓ This is an Answerable Question ✓

ANSWER

The idea of alien visitations in the distant past is an engaging concept that has provoked a vast amount of literature and discussion. However, the evidence that has been proposed as proof of such visits is easily dismissed, and most contemporary scholars give no credence to the notion. One of the difficulties is that someone makes an incorrect statement (for example, the Piri Reis map shows Antarctica, which it doesn't) and this gets reprinted and perpetuated, so that in due course the original error is perceived as established fact.

There are indeed many magnificent buildings and objects in existence that would have required exceptional skill and care in their creation. That may be a cause for wonder and appreciation, but it is far removed from evidence for extra-terrestrial assistance.

Our ancestors had the capacity for thought and experimentation. They also had the resources of vast labour forces. Every manufactured item so far discovered on our Earth is almost certainly of terrestrial origin. Until such time as we find irrefutable evidence of an extra-terrestrial artefact (such as an alien might consider one of our current travelling space probes), then we must assume that there have been no alien visitations to our planet. Accidental likenesses, such as the face on Mars or the Abydos helicopter, are just that – accidental likenesses. The evidence just isn't good enough to say that these were the creations of alien creatures however much we may wish that they were.

Question 7:
Alien Encounters – true or false?

Seeing what looks like a flying saucer is one thing but perhaps the most fantastic of all claims is that of actually meeting an alien. Such people are known as *contactees.*

There have been thousands of such claims many of which involve being abducted by aliens rather than just meeting them. Those who are abducted are known as *abductees.* Such reports date mostly from the 1950s onwards.

George Adamski

A well-known contactee was George Adamski (Fig.61), a Polish-American who published numerous UFO photographs, articles and books about his experiences.

Fig. 61 George Adamski (1891 – 1965)

In 1952, Adamski claimed to have had an encounter in the Californian desert with an alien from Venus who took him on a ride in their scout ship (Fig. 62) and then in their mother ship.

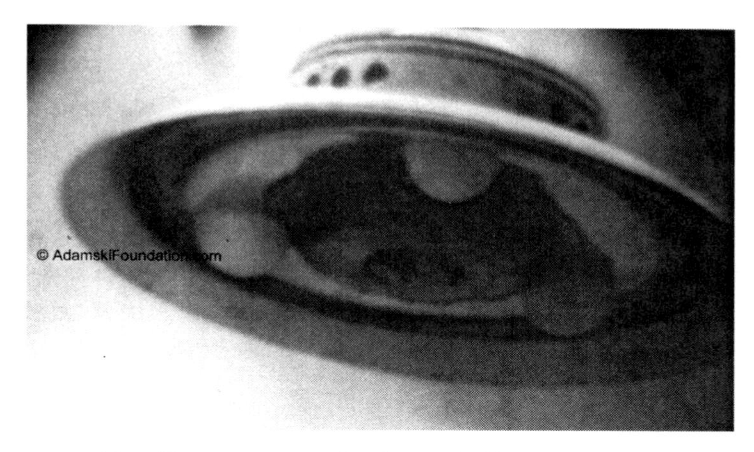

Fig. 62 Venusian scout ship photographed by George Adamski
©GAF International/Adamski Foundation, POBox 1722, Vista, CA92085, USA

Unfortunately, Adamski's camera was unable to operate satisfactorily inside either craft, the failure being blamed on high magnetic fields. Also, the Venusians would not allow him to take their photographs during their meeting in the desert. That of course was a great shame since his camera would presumably have worked perfectly well on the ground in California.

Adamski also recounted meetings with beings from Mars and from Saturn.

The space age hadn't started in 1952 so detailed knowledge about surface conditions on the planets was not available. Later information from space probes sent to Venus showed that the planet had a surface temperature of nearly 500°C (easily hot enough to melt lead), an atmospheric pressure over 90 times that on Earth (equivalent to a dive to over 3,000 feet which would require a full hard diving suit), and an atmosphere containing no oxygen.

The consensus opinion now is that no life could survive on the planet so how the remarkably human-like Venusians (Fig. 63) managed to exist under these conditions seems hard to explain.

Fig. 63 Adamski greeting his Venusian visitor in the desert. (from George Adamski, The Story of a UFO Contactee - chapter in How to Make the Most of a Flying Saucer Experience by Professor Solomon. Drawn by Steve Solomon. © Top Hat Press 1998)

Saturn is composed of gas and has no solid surface. It may be theoretically possible for some bizarre life forms to exist deep within the gas but this seems an unlikely habitat for a humanoid creature as described by Adamski.

Much like Erich von Daniken in the 1960s – see Question 6 – George Adamski must be given credit for his work which created a huge international following of UFO enthusiasts. He gave lectures around the World and published a number of successful books about his exploits. The best known is probably his first - *Flying Saucers Have Landed*, which was written with his friend Desmond Leslie and published in 1953 (Fig. 64). It provides a detailed account of his meetings with the aliens and includes several photographs and drawings. The book is out of print but a pdf file is available to download at www.universe-people.com

The excellent cover art shows the scout ship hovering over the Californian desert with the mother ship waiting in the background.

Fig. 64 Flying Saucers Have Landed

Adamski's fan base shrunk considerably when, in 1959, he announced that he was going to a conference on Saturn. Even for him, that was probably a giant leap too far.

BETTY AND BARNEY HILL

One of the best known alien abduction reports is that of Betty (1919 – 2004) and Barney Hill (1922 – 1969). On 19 September 1961, they were driving home through the White Mountain area of New Hampshire. They saw a white light in the sky and stopped the car. Barney got out and went to study the white light more closely with binoculars. He became convinced that he had seen a UFO.

Both described it as a pancake or banana-shaped object (Fig. 65). Barney stated that he could see beings behind the craft's windows. Soon after they returned to the car and drove home.

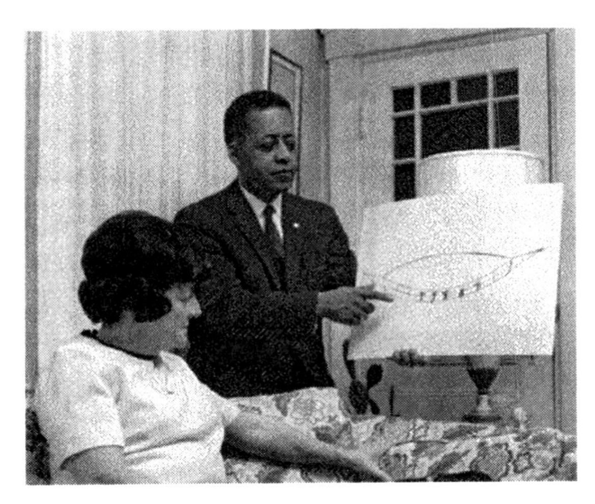

Fig. 65 Barney and Betty Hill with a drawing of the space craft in which they claimed to have been abducted. © Barney and Betty Hill

The couple reported their experience to the nearby Air Force Base and also to a national UFO group.

Both the Hills began suffering from a variety of ailments. Betty had a series of nightmares involving beings with cat-like eyes and Barney had a painful back. They could also not account for 2 hours of their time during their journey home.

In 1964, 3 years after the event, they were referred to a psychiatrist who recommended a session of hypnotic regressions. Under hypnosis, the Hills revealed that the UFO had landed close to the car and they were somehow partially sedated. They were then forcibly taken aboard the space craft by beings with large cat-like eyes and white skin. Inside the craft, the aliens carried out various medical examinations and some tissue samples were taken.

Finally, the couple were taken on a tour of the space craft. Betty asked whether she could take a book as evidence of the encounter but this was denied. She asked where they came from and was shown a star map and recalled its details under hypnosis.

The Hill case became worldwide news and spawned a book *The Interrupted Journey* by John Fuller as well as a 1976 made-for-TV movie *The UFO Incident.*

It would be easy to dismiss this case as a fantasy brought on perhaps by tiredness and the sighting of a military aircraft from the nearby Pease Air Base. Easy, except for one thing. Under hypnosis, Betty Hill re-created the star map she had been shown by the aliens (Fig. 66).

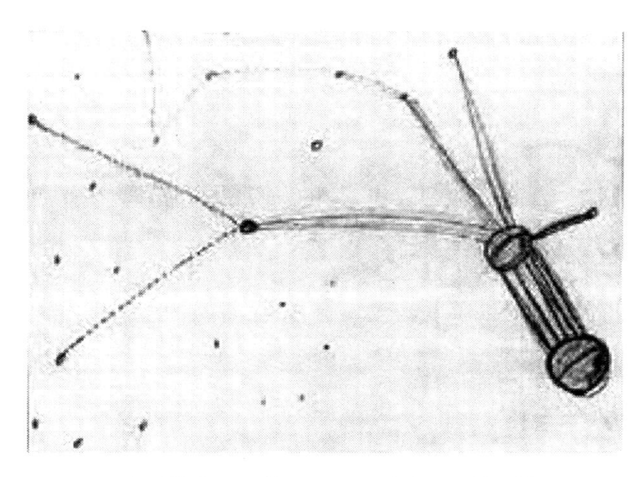

Fig. 66 Betty Hill's star map. © Betty Hill 1964

Two years later, in 1966, Marjorie Fish, a school teacher and amateur astronomer, wondered whether the Hill star map bore any resemblance to an actual star pattern. Her eventual interpretation is shown in Fig. 67.

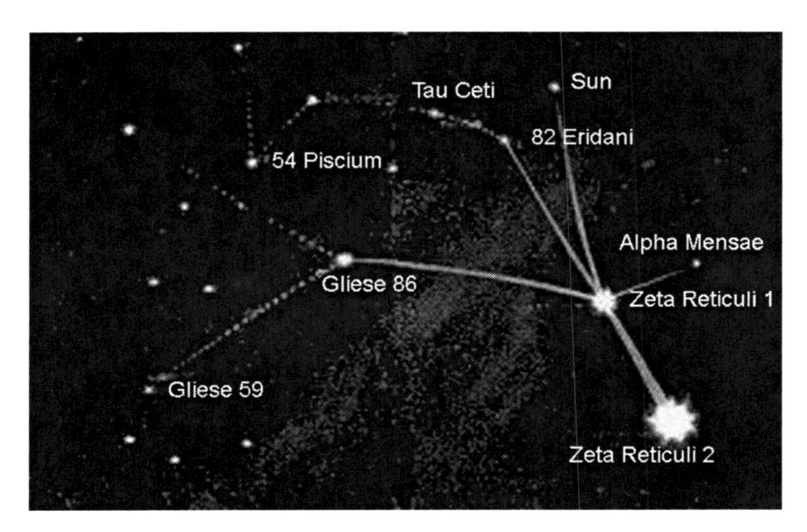

Fig. 67 Marjorie Fish's interpretation of the Hill star map. © Marjorie Fish 1966

The similarities are uncanny and led many to believe that the aliens encountered by the Hills came from the binary star system *Zeta Reticuli*. It is particularly impressive when one considers that some of the stars in the drawing were unknown until the Gliese Star catalogue of 1969 was published. It should also be mentioned that currently (2011) no planets have been detected in the Zeta Reticuli system (although of course that doesn't mean they do not exist).

So what is the explanation? How did Betty Hill manage to draw an apparently accurate map of a distant star system when some of the constituent stars were not even known at the time?

As always, we have more than one choice and it is prudent to heed William of Ockham's advice and go for the simplest explanation that makes the fewest assumptions.

An excellent analysis is given in the December 1974 issue of *Astronomy Magazine* in an article entitled *The Zeta Reticuli Incident* by Terence Dickinson. This included a section by Carl Sagan and Steven Soter in which they gave a detailed critique of Fish's interpretation. Their main point was that both the Hill drawing and the Fish interpretation looked similar because of the lines connecting selected stars. Remove the lines and replace some of the stars that had been left out of the published drawings and the similarity disappears. Rejoin different stars with different lines and the similarity also disappears. The basic idea is illustrated in Figs. 68 and 69.

A star field is a selected region of the sky that might contain many hundreds or even thousands of stars. To keep it simple, there are 15 stars represented in Fig. 68A and in Fig. 68B. They represent different star fields, that is, different regions of the sky and not surprisingly, the two drawings look completely different.

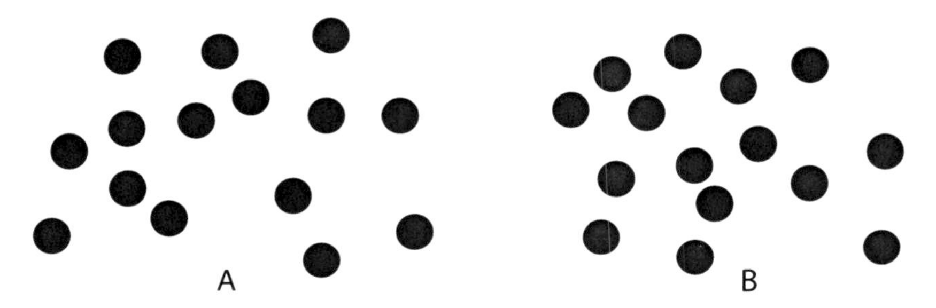

Fig. 68 Two random collections of dots (A and B) representing two star fields

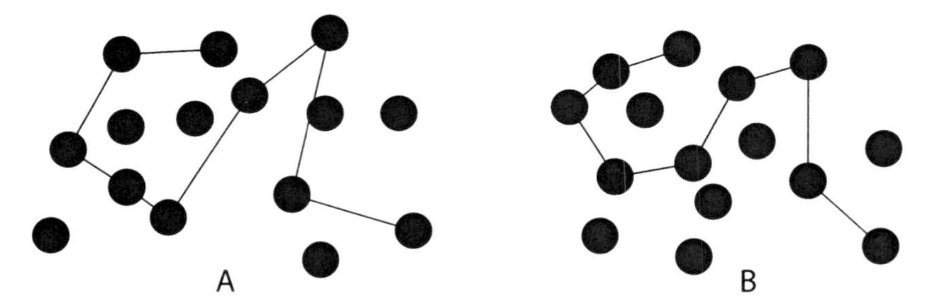

Fig. 69 The same star fields shown in Fig. 68 with selected stars joined together by lines

Now look at Fig. 69. Stars have been selected in Fig 69A and in Fig. 69B so that the joining lines create similar patterns. This is possibly why the Fish interpretation of the Hill map (Figs. 67 and 66) was such an impressive likeness. With a large number of stars to select from, it would be a simple matter to join just those that repeated Betty Hill's original pattern.

To make the comparison even more persuasive, it would have also been easy to remove many of the superfluous stars that only appeared in one of the illustrations. The result would be a good reproduction of the Hill pattern and with many of the unconnected extra stars in similar positions.

So which is the more reasonable explanation? If you want to believe in UFOs and aliens, then you would choose the Hills' version but if you prefer to make your decision based on credible evidence and what seems reasonable, then you might chose Professor Sagan's version.

There have been many more reports of human-alien encounters but these will suffice as a fair sample.

✓ THIS IS AN ANSWERABLE QUESTION ✓

ANSWER

There may indeed be aliens cruising around the galaxy in space vehicles but until we have irrefutable evidence of such beings we must remain sceptical. Anecdotal evidence is of little value; photographs could be faked even in the 1950s; drawings may or may not represent what was seen; sightings of purported alien space craft are more likely to be of military aircraft, perhaps secret and experimental ones; and apparently scientific analyses may be open to alternative and more likely explanations.

Based on what we currently know our answer is that reports of alien encounters are untrue either due to deliberate falsification or unintentional hallucinations or false memories.

Question 8:
Is there extra-terrestrial intelligent life?

The question of whether there is intelligent life anywhere else in the Universe is a fascinating one. Can we answer this question, or at least make a Best Guess? (What do we mean by 'intelligent life'? All life is intelligent in the sense that it knows what to do in order to keep itself alive and to reproduce. Intelligent life in the present context though is usually taken to mean life with which we could, in theory, communicate through the medium of radio transmissions).

There are three possibilities.

1. Yes, there is intelligent life elsewhere in the Universe

2. No, there is no intelligent life elsewhere in the Universe

3. There was intelligent life elsewhere in the Universe in the past but not now (parent star exploded; the life became extinct; the life was wiped out by some cataclysmic event)

The basic argument in favour of extra-terrestrial life is simple. With billions of stars and billions of galaxies, there must be intelligent life somewhere else as well as on Earth. If that was true, how could we communicate with each other? The obvious answer is by radio.

COMMUNICATION WITH EXTRA-TERRESTRIAL CIVILISATIONS

The first primitive *radio telescopes*, devices which could receive radio signals from space, were produced in the 1930s. The rapid development of radar during the Second World War then stimulated the formation of a new field of research – *radio astronomy.*

Professor Frank Drake, an American astronomer and astrophysicist, had a deep interest in the subject of extra-terrestrial life and how we might be able to communicate with it, and founded the organisation known as SETI, the *Search for Extra Terrestrial Intelligence.* This was set up in Green Bank, West Virginia, USA in 1961 and its radio telescope began to listen for transmissions from space in the hope of detecting messages from alien civilisations. SETI's website at www.seti.org has a number of interesting articles and a Frequently Asked Questions section, and is well worth visiting.

Professor Drake devised an equation, known as the *Drake equation,* to try and work out how many such civilisations existed in our galaxy with which we might be able to communicate.

Drake's pioneering idea was to consider those factors which were likely to be important in the development of intelligent life, and to put these together in an equation. Here's what he came up with (Fig. 70).

$$\mathbf{N = R \times f_p \times n_e \times f_l \times f_i \times f_c \times L}$$

Fig. 70 The Drake Equation

This is what the various symbols mean:

N is the number of civilisations in our galaxy which we which we could communicate. This is the number we are trying to calculate.

R is the number of new stars of the same type as our Sun that are formed annually in our galaxy.

$\mathbf{f}_p$ is the percentage of Sun-like stars that have planets.

$\mathbf{n}_e$ is the average number of planets, for every Sun-like star that has planets, that could support life.

$\mathbf{f}_l$ is the percentage of planets out of those that could support life, that actually have developed life.

$\mathbf{f}_i$ is the percentage of planets with life that go on to develop intelligent life.

$\mathbf{f}_c$ is the percentage of intelligent life forms that use communication technologies compatible with ours, that is, electromagnetic transmissions such as radio waves.

L is the average amount of time, in years, that such a civilisation exists before becoming extinct or being destroyed.

This was an innovative concept and was the first serious attempt to quantify the likelihood of making contact with extra-terrestrial civilisations. Professor Drake's estimated values, in 1961, for the various components of his equation gave a value for N = 10 although the figures have been revised many times to take account of the latest information.

There is however a really big problem with this equation that, in many people's views, renders it meaningless.

Although it is possible to make educated guesses, based on actual research data, for some of the values in the equation, most are utterly unknown and unknowable. There is absolutely no information at all for the last five components of the equation. It's a pure guess based on your own gut feeling.

You could put in any value at all for these components and they would be just as valid as any other. Your guess is as good as anyone else's.

The problem is exacerbated because the factors have to multiplied together. Let's assume that the current estimates for the first two components are reasonably accurate. Let's also be very generous and assume that the values for the last five components are in error by a factor of just 10. That means that the value for N is in error by a factor of 10 x 10 x 10 x 10 x 10 = **100,000**. However, if the guessed values for the last five components are in error by a factor of 100, then the final result is wrong by a factor of **10 billion!**

The trouble is that with just one example of life having formed (on Earth), we can't make any assumptions about how likely it was. If we should find life somewhere else, Mars maybe, or one of the moons of Jupiter, then the whole situation changes and we could justifiably say that life is likely to be quite common under the right conditions. So until we know more about the properties of extra-solar planets, and until we find evidence of extra-terrestrial life, the Drake equation remains unusable.

There is however an intriguing other possibility that might produce results far quicker than waiting many decades for the necessary evidence from other moons and planets to fit into the Drake equation.

The basic biochemistry of all life on Earth, whether viruses, bacteria, plants, trees, fish and mammals, is very similar. It is based on nucleic acids, RNA and DNA, to pass genetic information on to offspring, and it depends on proteins and fats and carbohydrates to keep itself alive and well. This has led to the currently accepted view that all life on Earth has a single common ancestor from which everything else developed. Put another way, life started on Earth just once.

Question 3 – *How did Life Begin?* explained the concept of molecules that are either left or right handed, like a pair of gloves. The scientific term for this handedness is *chirality*. Supporting evidence for a single common ancestor is that all sugars found in living creatures are of the ***d*** (right handed) form, and all proteins are made of amino acids of the ***l*** (left handed) form. If life had started several times, then it would be reasonable to expect some of it to have a different biochemistry, one version of which could be a different chirality, that is, it might be using left handed sugars or right handed amino acids.

Such life forms would look the same as their other-handed counterparts but they would be completely incompatible with them. An apple, for example, based on a reverse chirality biochemistry, would be indigestible and therefore completely useless as food for a creature with a biochemistry based on normal chirality. You could play with it, but you couldn't eat it. (Well, you *could* eat it but it would give you zero nourishment).

Left and right handed gloves are a good visual example of opposite chirality.

All life so far found on Earth has ***d*** sugars (right handed) in its carbohydrates and ***l*** amino acids (left handed) in its proteins. Other alternative biochemistries could possibly exist, and some research workers are looking for them. A potential candidate is a species of bacteria known as GFA-J1 found in Mono Lake, California, USA.

This organism seems to be able to use arsenic in place of phosphorus in its DNA and in other essential biochemical molecules. However, the lake is very rich in arsenic and it needs to be confirmed that this element is really a component of GFA-J1's molecules and not just a contaminant. Even if this finding is confirmed however, it is still some way from being accepted as a separate biochemistry since the basic molecules are still the same as those used by the rest of the creatures on Earth.

If even one example of an organism with a different mechanism for passing genetic material, or with all of its proteins constructed of right handed amino acids, then that would be good evidence that life had arisen on Earth more than once. If it's happened more than once here, then it could also have happened elsewhere. We can then make a better guess at some of the components of the Drake equation.

For now though, all we know for sure is that life has arisen once, and any results from the Drake equation must, for now, be considered invalid.

SETI has been searching for extra-terrestrial life for over 50 years and so far there has been no irrefutable evidence for any radio signals that could be construed as a message. The so-called *Wow! signal* (Fig. 80) was a one-off event and in the absence of any repeat signals has to be considered as a technical rather than an extra-terrestrial event.

The probability of extra-terrestrial life

As stated at the beginning of this topic, the usual argument in favour of extra-terrestrial life is straightforward. With billions of stars and billions of galaxies, there must be intelligent life somewhere else as well as on Earth. We can't work out the probabilities since, as we've seen in the discussion of the Drake equation, we don't have sufficient information. But we can put it into some sort of context.

We need to make an assumption, which is that the life we are talking about is such that it has to arise on a rocky planet revolving around a star. We might be able to conceive of some bizarre life forms that could exist in space, or inside stars, but as we know absolutely nothing about such theoretical life forms, we are not in a position to consider them.

The problem is that it's impossible to work out the probability of something happening if it's only been observed once. Take planets. It's been known since the invention of the telescope in the 1600s that the planet Saturn had rings round it.

When Galileo first looked at Saturn through his newly invented telescope in 1610, he was astonished to discover that it seemed to consist of three bodies.

It wasn't until Christian Huygens looked at Saturn with his improved telescope in 1655 that the rings were discovered (Fig. 71) although Huygens' view wasn't quite as good as this!

Fig. 71 Saturn and its rings photographed by Voyager 2 in 1981. © NASA

Let's pose a question. Let's assume we're in the pre-space age, and wonder how many planets have rings round them? Of the nine planets in our Solar System (actually only 8 now since Pluto was downgraded to dwarf planet status by the International Astronomical Union in 2006), only one has rings. So what proportion of planets have rings? In our Solar System, it's one out of 9, that is, just over 11%.

But what about any planets that might be discovered in other Solar Systems? Is the percentage of planets with rings still about 11%?

Who knows? We just can't say. It might be more or it might be less. Since we've only observed it once, we don't know if ring systems are a very rare event that just happened to form around Saturn, or if they're very common and would form round many planets, even though it only happened once in our system.

Information from spacecraft and improved earth-bound telescopes now enables us to answer our question. In 1977, two American astronomers using an airborne observatory, basically a telescope in a plane, discovered rings around Uranus.

In 1979, the Voyager 1 spacecraft discovered rings around Jupiter, and in 1989, Voyager 2 discovered rings around Neptune.

All the giant gas planets in our Solar System, Jupiter, Saturn, Uranus, and Neptune therefore have rings around them. It's a fair bet then that ring formation is a common occurrence around large planets but not around small rocky ones like Mercury, Venus, Earth, and Mars (and Pluto).

Now that we know that all the other gas giant planets in our Solar system have rings, we can say with some confidence that planetary ring systems around gas giant planets are likely to be common. When we only had Saturn as an example of a ring system however, we had no idea whether they were common or rare. It's the same with Life – one example tells us nothing about how common it is.

Earth-bound telescopes, the Hubble space telescope, and various space craft have made extensive studies of every planet and many moons in our Solar System. To date there has been no evidence of any life, intelligent or not. The most detailed studies have been on Mars where robot landers have been surveying the landscape and digging up the soil and analysing it since the 1970s (Fig. 72).

Fig. 72 Martian surface from NASA's Spirit Rover, 2004. © NASA

Some more recent photographs of the Martian surface have shown what could be dried up rivers and lakes, leading some people to suggest that there may have been liquid water on the surface at some time in the past. The Jovian moon Europa seems to be covered with a thick layer of water ice which has led some astronomers to speculate that there may be liquid water beneath. There's even some evidence that water may have been present on our own Moon in the past.

Life depends on chemical reactions. Living creatures need to feed, grow, repair themselves, reproduce, and interact with their environment. All this needs chemistry, and chemistry has to take place in some type of support medium. For all of us on Earth it's water.

The absence of liquid water on other planets and moons makes it unlikely that life has developed there. Not impossible of course because other substances such as liquid ammonia or hydrogen sulphide could replace water as the medium.

It's easier however to start with what we know, so the presence of water on other planets and moons is an exciting discovery since it at least makes it possible that some sort of life may have developed there.

Earth is a small rocky planet. So are Mercury, Venus, and Mars. And so is Pluto although it's now no longer classified as a planet. Space probes have visited all except Pluto, which is due to encounter NASA's New Horizons craft on 14 July 2015.

There has been no evidence of life on any of these planets. They are either too close to the Sun and therefore too hot, or too far away and therefore too cold. As it turns out, the Earth is in just the right position being neither too hot nor too cold for life to survive. Had it been a bit nearer to the Sun or a bit further away, we wouldn't be here now. Life can only exist within a very narrow range of extremes.

This applies to many things. A blood test print out will show the amounts of a whole variety of substances in your blood, including minerals, enzymes, glucose, blood cells and fats. You'll notice that next to each actual result, there will be a range of values. These represent what doctors call the *normal range*. If your result is within this range, then it's OK. If it's too high or too low, then that's an indicator that something's wrong.

Urea and electrolytes			
Serum sodium		140 mmol/L	(137-145)
Serum potassium	HI	5.5 mmol/L	(3.6-5.0)
Serum urea level		6.2 mmol/L	(3.2-7.1)
Serum creatinine		88 umol/L	(58-110)

Fig. 73 Blood test print-out for urea and electrolytes showing an abnormally high value for potassium

These ranges are quite narrow. A normal range for the amount of potassium in the blood for example, is between 3.6 and 5.0 units. The result shown for potassium in Fig. 73, a value of 5.5, has been automatically flagged as high on the print-out.

To date, we have found no evidence whatsoever for any alien life forms in our own Solar System. Having said this however, mention should be made of the discovery of some unusual chemistry on the surface of Saturn's moon Titan.

Titan is unique in being the only moon in the Solar System with an atmosphere. Over 98% is nitrogen with the rest being mainly methane and hydrogen. Data from the Voyager and the more recent Huygens and Cassini space probes have found surface lakes of liquid methane and ethane.

In 2005, two NASA scientists speculated that an alien life form could perhaps exist in these lakes even though the surface temperature is around -180°C. They theorised that such life would consume hydrogen from the atmosphere and acetylene, which would be solid at Titanian temperatures, from the moon's surface.

This could produce energy for the organism. However, such a chemical reaction would need special conditions and/or special catalysts to proceed, much like happens with biochemical reactions inside terrestrial organisms. If such reactions were indeed taking place on the surface of Titan (Fig. 74), and in 2010 new measurements indicated that this may be the case, then this might indicate the presence of living organisms.

On the other hand, it might also just indicate the presence of some novel chemistry on an alien world.

Fig. 74 Surface of Titan taken by Huygens probe. ©NASA 2005

A more sophisticated Titanian lander would be able to make direct measurements on the surface of the moon, and would also be able to sample surface material to look for living organisms. This however is unlikely to happen for several decades, if it happens at all. Meanwhile, Titan's novel surface chemistry must remain an interesting enigma.

Mars, which depending on the relative positions, is sometimes the closest planet to Earth, has been extensively studied for signs of life. Although nothing has been found so far, the *ExoMars* mission planned for 2019 will photograph, drill into and analyse samples taken from below the Martian surface. If anything is or has been living there, it is hoped that the *ExoMars* rover will find it. We'll have to wait and see.

Our search for extra-terrestrial life now moves on to the possibility of other Solar Systems, that is, planets orbiting other stars. We have planets orbiting our star, the Sun, but how common are planets around other stars? Again, with just one example, our own star, we couldn't say. It may be very common or exceedingly rare.

This all changed in 1992 with the discovery of the first exo- (outside our Solar System) planet. NASA's Kepler spacecraft, launched in 2009, is designed to search for extra-solar planets by measuring the slight loss of starlight as a planet passes between the star and the observer, in this case the Kepler telescope. This is known as a *transit.*

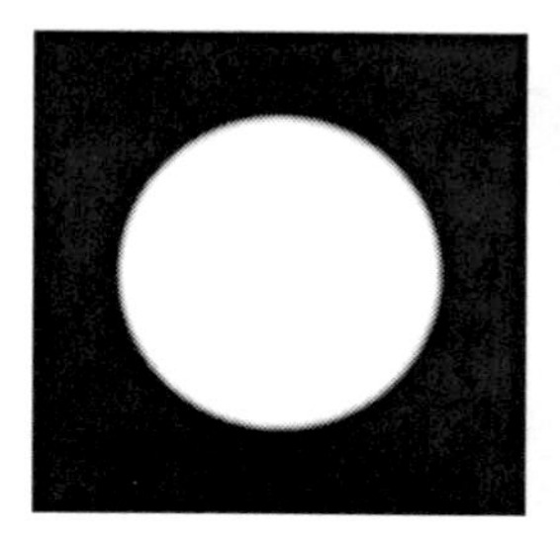

Fig. 75

Fig. 76

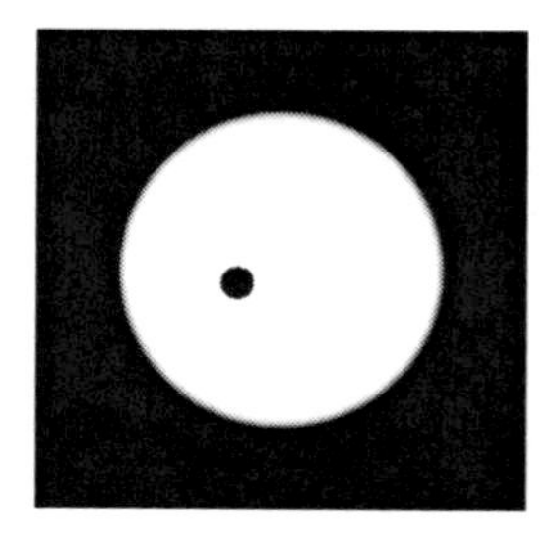

Fig. 77

Transit of a star by a planet

Fig. 75 shows a star which emits a certain amount of light that can be measured by instruments attached to a telescope. If this star has a planet going round it, then when the planet comes round from the back of the star it becomes visible against the backdrop of the star (Fig. 76). It remains visible as it crosses the face of the star (Fig. 77) and eventually disappears from view as it once again goes behind the star.

It is clear that the amount of light measured as coming from the star in Fig. 75 is greater than that in Fig. 76 and Fig. 77, since some of the light will have been blocked out by the planet. This passage of a planet in front of a star is called a *transit* and the Kepler telescope will be looking for such transits.

Not only can new planets be detected by this process, it is also possible to determine the size of the planet, how long it takes to orbit its star, and how far it is from its star.

In February 2011, NASA announced some results from the Kepler mission which showed that 1,235 exo-planets had been discovered. These planets varied in size from Earth-size (68 planets) up to much larger ones, some even exceeding the size of our largest planet, Jupiter. Most interestingly though, 54 of the new planets were located in the *habitable zone* around their star, a zone where liquid water, and perhaps life, could exist. Of those 54 planets, 5 are similar in size to the Earth.

So we can safely say that planets are common, and that some stars have Earth-sized planets circling in the habitable zone. It does seem likely then that our galaxy, and presumably all other galaxies, are full of planets, and that some of these will be small and rocky and at the right distance from their star for liquid water to exist. What about life?

We can only discuss life as we know it although that covers a wide enough variety. Bacteria, amoebae, sea anemones, flies, worms, trees, flowers, sparrows, alligators, horses, humans, etc. We can't discuss imaginary life forms such as may exist inside the Sun or floating free in space since we have no concept of what form they may take. However, it is certainly true that life can exist under a large variety of conditions, so although most life thrives within a narrow band of temperature, pressure and acidity, many more primitive life forms have been found in bizarre environments on Earth. They are sometimes called *extremophiles* (Fig. 78).

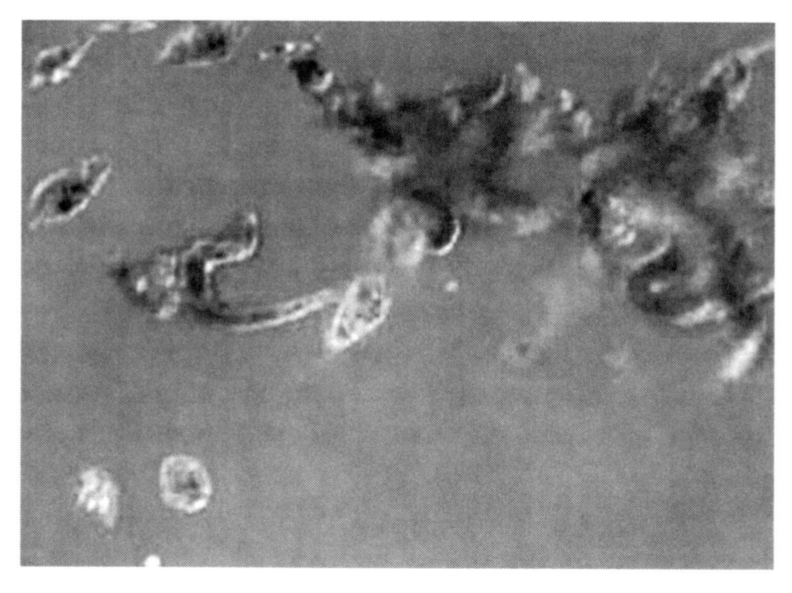

Fig. 78 Single-celled organisms found inside Berkeley Pit, Montana, USA from Wikimedia Commons (original @ http://toddtrigsted.com)

Berkeley Pit is a disused copper mine in Montana, USA. It is about half full of very acidic water with high concentrations of heavy metals such as arsenic, cadmium and zinc. In spite of this, new species of bacteria and fungi have been found to grow in these conditions; some of them actually ingest metals as part of their biochemistry.

Most extremophiles are simple single-celled organisms that have adapted to life under extreme conditions. Although there are a few more complex life forms such as worms and fish that also exist in such environments, no complex creatures such as mammals have been found with this capability.

In spite of these strange organisms therefore, most life needs to be within quite a narrow band of values. So our potential life-containing extra-solar planet needs to be within what has been called the *habitable zone* or the *Goldilocks Zone (neither too hot nor too cold but just right)* round its parent star. This means that liquid water can exist (essential for life since a liquid medium is needed to bathe living cells with nutrients and other vital materials).

We could perhaps conceive of a life form composed of solid crystals, or bathed in liquid nitrogen at -196°C, or in molten glass at 1,000 °C but this is a pointless exercise until there is at least some evidence that these could exist.

So our extra-solar planet needs to be the right distance from its star to maintain an appropriate surface temperature. We have no idea how likely this would be.

It also needs to contain the basic chemicals for life, not necessarily the same as ours of course, to form and we also have no idea how likely this would be. We therefore have three imponderables.

I) DO PLANETS FORM AT SPECIFIC DISTANCES FROM THEIR STAR?

Bode's Law, named after the German astronomer Johann Bode and published in 1778, provided a mathematical formula for the average distances of the planets from the Sun.

It worked for all the then known planets; it correctly predicted the existence of a large asteroid, Ceres, between the orbits of Mars and Jupiter, and it also correctly predicted the presence of Uranus, discovered 3 years later in 1781. However, it failed with Neptune and Pluto.

Most astronomers now consider Bode's Law as being meaningless although its partial success has not been properly explained. Unfortunately therefore we have no way of predicting star to planet distances and cannot say how likely it is that a planet would form within the habitable zone of its star.

II) HOW LIKELY IS IT THAT PLANETS CONTAIN THE CHEMICALS NECESSARY FOR LIFE?

The six most important elements needed for life (on Earth) are amongst the most abundant elements in the Universe:

LIFE	UNIVERSE
Hydrogen	Hydrogen
	Helium
Oxygen	Oxygen
Carbon	Carbon
	Neon
Iron	Iron
Nitrogen	Nitrogen
	Silicon
Sulphur	Sulphur

This is encouraging. It seems reasonable therefore to conclude that the basic building blocks needed for life would be present on a typical planet.

iii) Given the right planet at the right distance with the right ingredients, how likely is it that life will arise?

This is the big question that we struggle to answer. It's happened once to be sure, but we have no clues whatsoever as to how likely it was. All we know is that it took about 1 billion years. (It is f_l in the Drake equation).

So let's now try and make some calculations. Current estimates indicate that there are perhaps 100 billion (10^{11}) galaxies in the Universe, and that a typical galaxy may contain about 500 billion stars (5×10^{11}). This gives us a figure of 5×10^{22} for the total number of stars in the Universe.

Stars come in different types ranging from supergiants to white dwarfs. Giant stars and small dwarf stars may not be life-sustaining (wrong type of radiation; too short a life span and other reasons), and only about 5 percent of all stars are in the same class (spectral type G) as our Sun. So out of the total number of stars, there may be about 2.5×10^{21} that are similar to our Sun.

How many of these stars will have a small rocky planet located in the habitable zone that contains the ingredients and atmosphere needed for life to develop? Who knows, but until we have more data from the Kepler mission, let's be generous and say that they all will. This then gives an upper limit of **2.5 x 10^{21}** such potential life-sustaining planets in the known Universe. On how many of these planets will life actually begin to evolve?

We know of one, but have no evidence for any others. It may be very common, or it may be very rare, or indeed it may be unique to Earth. At our current state of knowledge we have no way of knowing, and this is the problem with trying to solve the Drake equation (Fig. 70).

As discussed in Question 3, a succession of events, that is, a process, had to occur for life to begin on Earth. Such a process, while unlikely to be the same as that which took place on Earth, would also have to occur on an alien planet for life to begin.

Now consider this. Let's replace the process that led to the origin of life on Earth with an imaginary robot. Imagine also that there is such a robot on each of the 2.5 x 10^{21} potential life-sustaining planets in the Universe and that these robots have been programmed to toss a coin and to record the results, either heads or tails. It's a fair coin and a fair toss.

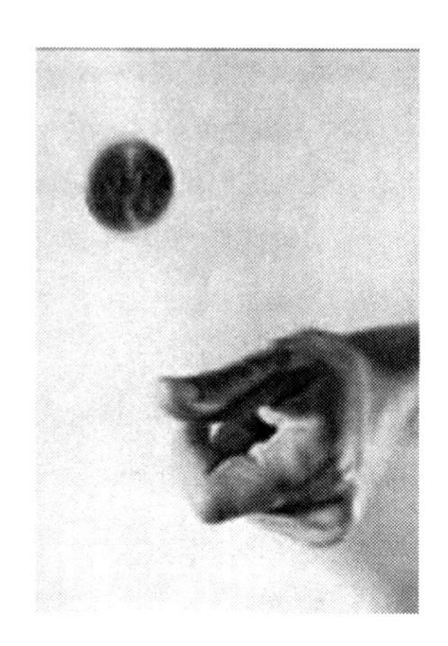

What are the chances of a coin coming up heads 71 times in a row? It's obviously possible but it's very unlikely. Actually, the chances are **1 in** $\mathbf{2^{71}}$ which is equal to about **1 in** $\mathbf{2.5 \times 10^{21}}$. (This is the same as the calculated number of potential life-sustaining planets in the Universe – see above - which is why the number 71 was chosen).

In other words, somewhere in the Universe, on one of the planets circling one of the stars in one of the galaxies, it's likely that one of the robots would have recorded 71 consecutive heads.

Tossing a coin is a very simple process since there are only two outcomes. It's easy to record the results and it's easy to work out the probabilities of different outcomes. In contrast, the origin of Life is a very complex process of which we understand very little. It's easy enough to record the result – a living organism – but much harder to work out the probabilities. In fact, it's impossible. We only know of one occurrence and that doesn't give us enough information to work out how likely it was.

Think of a die. Dice usually have 6 sides numbered from 1 to 6. The chance of rolling a 4 for example is then 1 in 6, just like the chance of a coin coming up heads is 1 in 2. One can however obtain dice with more or less than 6 sides as shown in Fig. 79.

Fig. 79 Dice with different numbers of sides
Left to right: 6 sides; 12 sides; 20 sides; 100 sides; 4 sides

It's easy enough to work out the odds of throwing any particular number with any of these dice. But imagine a die with an *unknown* number of sides numbered from 1 upwards.

Let's say we roll it and it comes up 4. What does that tell us about the chances of it coming up 4 a second time? Absolutely nothing.

We don't know how many sides it has so we can't work out the probabilities. The fact that it came up 4 on its first roll tells us nothing. If it came up 4 again on a second roll, then we'd have some history to work with.

It's the same with origin of life. All we know is that is has happened once. That only tells us that it is possible but it tells us nothing about how likely it was.

What if it was as likely as a coin coming up heads 71 times in a row? In that case, as we've seen with our imaginary robots, it would probably only have happened once in the entire Universe.

It's been said many times that with so many billions of stars and galaxies in the Universe, it's almost certain that many of them will have planets with life. We really aren't entitled to draw this conclusion since we don't know how likely it was.

Given enough time, life may develop on every suitable planet or it may be a chance event, like the consecutive head tosses, that while possible is so unlikely that it's probably only ever happened once.

The initial signs for extra-terrestrial life aren't encouraging. Marconi was awarded a patent for his radio system in 1896, and regular radio broadcasts from the BBC began in 1922. That was 90 years ago, so an alien living within about 45 light years of Earth would have had time to receive the remnants of these radio messages, realise that they came from an intelligent source, and send a reply.

Here is a list of stars within 45 light years of Earth that have planetary systems.

STAR	DISTANCE (light years)	PLANETS
Epsilon Eridani	10	2*
Gliese 876	15	4
Gliese 581	20	4
61 Virginis	28	3
55 Cancri	40	5
HD 69830	41	3
HD 40307	42	3
Upsilon Andromedae	44	4
47 Ursa Majoris	45	3

Taken from *List of Planetary Systems* in Wikipedia
*Unconfirmed

If any responses had been sent, then we should have started to receive them from about 1952 onwards.

Radio astronomers have been listening for signals from space that could have an intelligent source from the early 1960s, and continue to do so. That's right within the time frame for the planetary systems listed above, but so far nothing has been received that could be construed as coming from an alien life form.

This isn't conclusive evidence of course. Just because we've heard nothing doesn't mean there is nothing. There may be abundant life on some of these planets but none which has evolved sufficiently to build a radio telescope. All we can say so far is that we have no evidence of extra-terrestrial life.

However, we needn't actually restrict ourselves to stars within 45 light years.

Imagine that aliens from the Andromeda galaxy for example started sending out signals 2.5 million years ago. We'd be receiving these now, if they existed. The same applies to all of the other galaxies out there. Nor have we heard anything from those stars that have been targeted for close scrutiny by SETI investigators.

Mention should be made of what has become known as the *Wow!* signal. This was a strong radio signal detected by a SETI researcher, Dr Jerry Ehman, on 15 August 1977.

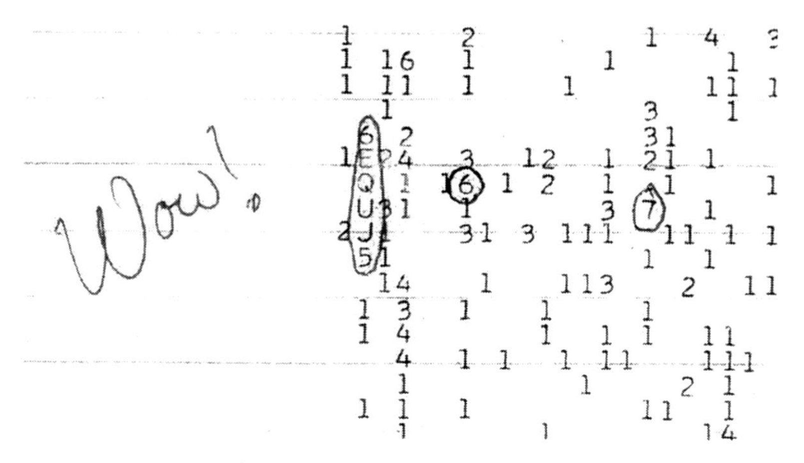

Fig. 80 Wow! signal print-out. © The Ohio State University Radio Observatory and the North American AstroPhysical Observatory

Fig. 80 shows the original computer print-out of the Wow! signal. The numbers and letters circled are a measure of the strength of the signal as compared to the background. It non-technical terms, the signal was 30 times stronger than any background noise, and lasted for the full time that the telescope was listening to that region of space, which was 72 seconds.

Unfortunately, in spite of repeated attempts since the signal was detected, it has never been heard again. So, much as many people would have liked this to have been a genuine part message from an alien civilisation, we have to assume that it was due to some other non-intelligent source. Even its discoverer cautions against making 'vast conclusions from half-vast data'.

This isn't conclusive evidence of course. Just because we've heard nothing doesn't mean there is nothing. There may be planets around some of these stars with abundant life but none which has yet evolved sufficiently to build a radio telescope. All we can say is that we have no evidence of any extra-terrestrial life.

This lack of evidence however hasn't stopped some scientists from making bold claims. In 2010, a team of scientists at the University of California announced the discovery of an Earth-sized rocky planet orbiting a star known as Gliese 581.

Gliese 581 is about 20 light years from Earth, and the new planet, Gliese 581g, orbits its star in the *Goldilocks* zone where liquid water could exist. This makes Gliese 581g the most Earth-like planet so far discovered.

The lead scientist, Professor Stephen Vogt, has stated, *"Personally, given the ubiquity and propensity of life to flourish wherever it can, my own personal feeling is that the chances of life on this planet are 100 percent."*

That seems like a very rash statement. It's a potential candidate for life but that's about as far as one can go. And although 20 light years may not sound very far away, it's actually about 120 trillion miles (120,000,000,000,000 miles). Even travelling at 1% of the speed of light, a currently unthinkable speed of about 7 million mph, a spacecraft would still need about 4,000 years for a round trip.

So, disappointing as it may be to some, there is currently no evidence at all for extra-terrestrial life. We may find it one day, or we may never find it because it isn't there – there's just us.

EXTRA-TERRESTRIAL LIFE DID EXIST BUT IS NOW EXTINCT

This is an interesting idea but unlikely ever to be provable. If the planet's parent star has exploded, then any evidence for life would have been destroyed. If the planet had suffered a major asteroid impact, or the life forms had become extinct or destroyed by some natural or artificial calamity, then remnants might remain to be discovered by future astronauts.

It's a bit like archaeologists coming across the remnants of an ancient civilisation, such as the ruins of Tiahuanaco, 12,700 feet above sea level in Bolivia (Fig. 81).

Fig. 81 Tiahuanaco ruins, Bolivia (circa 500 BCE)

Little is known about this lost civilisation since it left no written records. It probably arose in around 500 BCE and continued to thrive for 1,000 years or so after which it died out. So although we can't deduce much about the inhabitants from these ruins, we do at least know that someone was there.

But what if...

But what if we do one day we do actually receive an alien signal with irrefutable evidence that it is genuine.

There would be all manner of implications, not the least being that we would then know for sure that life was not a one-time occurrence. We could then reasonably speculate that life is likely to be common throughout the Universe.

An imaginary newspaper front page is shown in Fig. 82.

The Daily Star

Friday 10 October 2941 **Special Edition** **£175.00**

Alien message from Andromeda Galaxy

by Peter Altman
Staff Reporter

Andromeda Galaxy

The Andromeda Galaxy is 2.5 million light years from Earth and can just be seen with the naked eye. Three days ago, NASA's orbiting radio telescope received a series of transmissions from the direction of Andromeda in the form of a repeating series of binary pulses. No further information has been released by NASA except to confirm that the signals are unlikely to have come from a natural source.

Speculation is intense as to the content of the message although its decipherment could take some time. It has to be remembered however that the message would have been sent 2.5 million years ago so its senders will be long dead. Any reply that we might send back will then take another 2.5 million years to get there.

A two-way communication is therefore impossible but if the message is confirmed as authentic it will be the first irrefutable evidence that we are, or were, not alone in the Universe.

Even though most observers consider the announcement to be genuine, some are saying that the message has been fabricated to divert attention from secret military projects.

110101001001010101010
000010101010101011010
101010101010100000001
111101000110010101010
100001011100011001010

What the message might look like in binary form

Fig. 82 Imaginary newspaper report announcing receipt of alien signals from the Andromeda Galaxy

And what if aliens actually do *land* on Earth? Will they come out of their space craft and demand *"Take me to your leader!"* If so, we'll be ready because the United Nations have recently appointed a Malaysian astrophyscist named Mazlan Othman as their Space Ambassador for

Extraterrestrial Contact Affairs. Let's hope that she's not out when she gets the call.

✗ This is an Unanswerable Question ✗

BEST GUESS ANSWER

Life, as we must consider it, needs to arise on a rocky planet located within the habitable zone of its star and to contain the necessary ingredients for life. Planets seem to be common around other stars so it's likely that many stars have the right sort of planet in the right position for life to develop.

There are likely to be thousands of stars within a distance from Earth that gives enough time for an alien civilisation to have intercepted radio broadcasts from Earth and sent coded signals back to us. There is also the possibility of one-way signals sent millions of years ago from anywhere in the Universe to have reached us by now. There is no evidence whatsoever for any such signals having been received.

We have no information at all as to how likely it is that life will develop on a suitable planet and calculations attempting to quantify these probabilities are invalid due to the lack of suitable data. Life may be so unlikely that it has only happened once or it may be extremely likely so that it has happened many times. However, there is absolutely no current evidence for any intelligent or indeed any other life anywhere else in the

Universe. That doesn't mean it isn't there – it just means we haven't found it.

There is also no evidence that life has arisen on Earth more than once. Absence of evidence isn't evidence of absence but our Best Guess has to be based on what we actually know.

Our Best Guess answer therefore is that there is no evidence for intelligent life elsewhere in the Universe, and that its formation on Earth may have been an event so unlikely, like the putative coin tosses, that it has only happened once. This conclusion is the only one we are entitled to reach with present knowledge. The question of whether life had ever existed in the past but is now extinct is probably forever unanswerable.

Either we are alone in the Universe or we are not, and either result is equally profound.

Question 9:
Could we travel to other galaxies?

Will we ever travel to the stars, or to other galaxies? Could it be done?

It will take the New Horizons space craft, launched in 2006, nine years to reach Pluto, travelling at an average speed of about 37,000 mph. If the technology could be upgraded to take astronauts, then we'd be talking about a 20 year round trip, which is probably feasible (if anyone was prepared to be away that long).

But what about interstellar and intergalactic travel? The nearest star is over 4 light years away which is about 25,000,000,000,000 miles. Let's assume that new technology could increase the average speed of a space craft 100 times, taking it to nearly 4 million mph. Even at that speed, it would take over 700 years to reach Proxima Centauri. Another tenfold speed increase would still give a journey time of 70 years. An additional problem would be communication with mission control. As the craft neared its destination, 4 light years away, radio waves carrying messages would also take 4 years each way. Eight years is a long time to wait for an answer.

If we assume that a 20 year round trip for an astronaut crew is the maximum practical time for a space voyage, and that you wouldn't go all that way without staying for a while, we have to say that about 8 years for a one-way journey to Proxima Centauri is the longest time available. That means travelling at half the speed of light, over 300 million mph.

That's impossible with forseeable technology. There are various ideas on the drawing board however.

Space craft powered by ***nuclear pulse propulsion*** engines have been proposed and researched. These engines would develop millions of tons of thrust by having directed nuclear explosions driving the craft forwards. The aim was to design a space craft that could reach Proxima Centauri in less than a crew member's or mission control member's working lifetime (taken as 50 years).

Constant acceleration drives could also make inter-stellar journeys feasible. In contrast to current modes of rocket propulsion where there is an initial engine burst to get the space craft under way followed by the remainder of the journey in 'coasting' mode, a constant acceleration drive is on all the time, continuously accelerating the space craft. In other words, the space craft just goes faster and faster, eventually approaching the speed of light.

With a constant acceleration of 0.5***g*** (***g*** is the acceleration due to gravity on Earth, about 32 feet per second per second, that is, every second an object falls it goes 32 feet per second faster than during the previous second), a space craft would approach the speed of light in about 1 year. During this time, it would have travelled about half a light year in distance.

As an approximation, we can say that the journey time with such a constant acceleration drive would be the distance in light years to the destination plus 1 year. A journey to Proxima Centauri, just over 4 light years away, would therefore take about 5 years.

The effects of relativity will be relevant meaning that the elapsed time frame for the crew will be shorter than that for observers back on Earth. For example, a journey time from the crew's point of view of, say, 5 years might appear to observers on Earth as having taken 50 years.

The time difference would be much greater for longer journeys and could run into thousands of years.

The crew will in effect have travelled into the future. If they stay for a year and then return to Earth, they will have been away for 11 years from their point of view. On Earth however, 101 years will have passed. So, if they start their journey in 2050 and return 11 of their years later, it won't be 2061 but 2151.

Journeys to destinations that are significantly further away seem unimaginable even with nuclear powered constant acceleration drives. However, if the crew were to live and die on the space craft as families and future generations completed the mission and returned to Earth, or the crew remained in deep suspended animation for the duration of the journey, then in theory at least it could perhaps be done for the closer destinations within our galaxy.

It has been suggested that a mission which cannot reach its destination within 50 years should not be started at all. Instead, the money should be invested in designing a better propulsion system. This is because a slow spacecraft would probably be passed by another mission sent later with more advanced propulsion, making the first mission a waste of resources.

Inter- and intra-galactic travel with journey times measured in millions of years remains in the realms of science fiction. Such journeys will require more than just faster rocket motors. Distances measured in thousands and millions of light years could never be accomplished by straight line travel even at speeds close to that of light. Does that mean it will never be done?

There could be some theoretical ways around this.

TRAVEL AT FASTER THAN LIGHT SPEEDS

It is possible for certain things to travel faster than light. An easy one to visualise is the movement of a laser beam as the laser is moved. A beam is shone from Earth on to the Moon for example. The Moon has a diameter of 2,160 miles. If the laser is moved so that the beam traverses the Moon's diameter in one hundredth of a second, then the laser spot has travelled 2,160 miles in one hundredth of a second. This is equal to 216,000 miles a second which is faster than the speed of light.

Fig. 83 shows the set-up. L is the laser beam on Earth travelling along the path L1 to hit the Moon M at point M1. The laser is now moved very quickly to the left, in one hundredth of a second, so that its beam follows path L2 and hits the Moon at point M2. The beam has therefore moved from M1 to M2, a distance of 2,160 miles, in one hundredth of a second. This equates to a speed of 216,000 miles per second which is faster than the speed of light. Had the laser been moved even faster, say in one thousandth of a second, then the beam would be moving across the Moon at over 2 million miles per second!

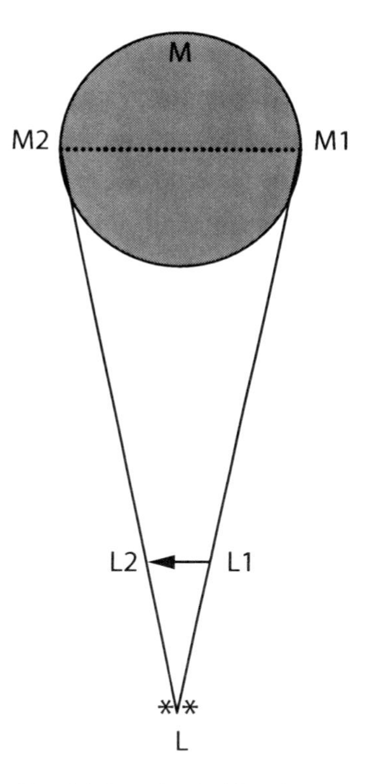

Fig. 83 Super-luminal laser beams

How can this be? Actually, no particle is travelling faster than light. A stream of photons (light particles) is hitting the Moon at M1. Then the beam begins to move sideways, indicated by dots on the path from M1 to M2. Each dot is composed of the next batch of photons to hit the Moon, and so on. Every photon sent by the laser from Earth travels at the speed of light as it hits the Moon. What is travelling at a faster-than-light speed is the laser spot, but this is composed of different photons each time it moves. So although the beam seems to be moving at super-luminal speeds, it is in reality a different beam made up of different photons each time it moves.

But what about material objects travelling faster than light? There are many entries on Google under the term 'faster than light travel'. Several of these are from scientists who have used advanced mathematics and physics to demonstrate, on paper at least, that this may be possible without infringing the known laws of physics.

If, for example, we could travel at one million times the speed of light, then we could go from one end of our galaxy to the other in about a month, and could reach the Andromeda galaxy in 2.5 years. It's a very long journey however from some equations in a research article to a working spacecraft.

There's another problem with such journeys. The fact that the Andromeda galaxy, for example, is 2.5 million light years away means that we are seeing it now as it was when the light left, that is, 2.5 million years ago. We don't even know if Andromeda is there *now*.

Even if one day we do develop a super-luminal (faster than light) transport system, we could arrive to find that it's gone – moved, exploded, who knows? OK guys, we're too late. Let's go back home.

TRAVEL THROUGH *WORM HOLES* IN SPACE

A *worm hole* is a purely theoretical concept. Imagine space as the surface of a balloon. To travel from one point on the surface of the balloon to another, you have to make a curved path. Some physicists have suggested that space is curved like this. If that's so, then a worm hole is like a path taken through the ball rather than over its surface. This path is of course shorter than the surface one so the journey time would be faster. As the curvature becomes more extreme, so the travel distance becomes shorter (Fig. 84). But it's only a theory.

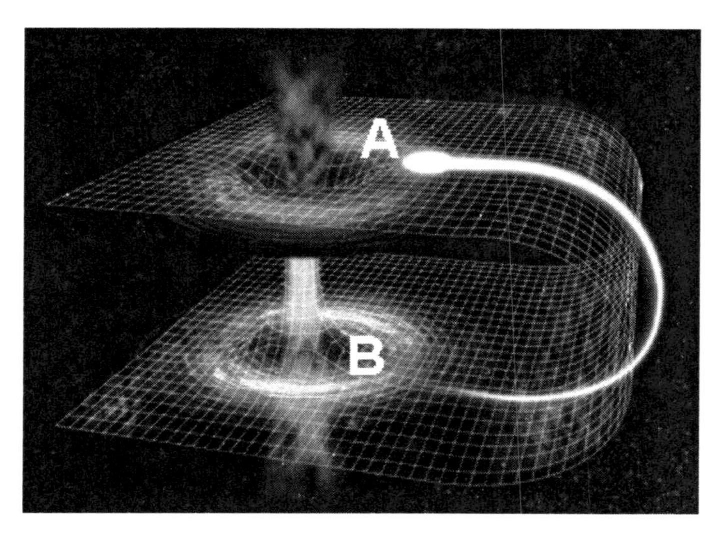

Fig. 84 Wormhole. Space is represented as a curved surface. A space craft going from A to B along the curve would clearly have a longer journey in both time and distance than a space craft going through the wormhole. © edobric-Fotolia.com

It's a brave person who says that something is impossible. Look at all the technology available today and imagine what people living 100 years ago would think if they could see it now. It's a mistake however to think that everything develops and will eventually happen. Here's an example.

The first controlled flight of a powered aeroplane was made by Orville Wright at Kitty Hawk, North Carolina, USA, on 17 December 1903 (Fig. 85) and here are the statistics of that flight.

Distance travelled 120 feet	Flight time 12 seconds
Speed 7 mph	Height 20 feet

Fig. 85 Orville Wright's first flight with Wilbur alongside (1903)
© NASA

Seventy three years of development resulted in the first flight of Concorde (Fig. 86). This aircraft had a range of 4,500 miles, could reach twice the speed of sound and cruise at 60,000 feet, high enough to see the Earth's curvature.

Fig. 86 Concorde

It doesn't always happen like that though. In 1969, Apollo 11 landed on the Moon (Fig. 87).

Fig. 87 Buzz Aldrin stepping on to the Moon's surface from Apollo 11 in 1969. © NASA

What would have been the expectation, in 1969, of how manned space exploration would develop over the next 40 years? Manned colonies on the Moon? Regular trips to Mars? And what's happened? Not much.

Why? The main reason is cost. Even Concorde was so expensive to develop (about £20 billion in today's money) that it became a shared project between the UK and France. The Apollo space program cost, in today's money, close to $200 billion. Public support is essential when Governments wish to spend money at that level, and after seven Apollo missions to the Moon, that support disappeared.

More complex projects, such as manned planetary missions and lunar colonisation missions, would be vastly more expensive, and with more immediate problems on Earth to deal with, the resources to develop these space missions were just not available.

However, in 2010 American President Obama announced his plan for a manned mission to Mars in the 2030s, starting with an orbital flight to be followed later by a landing. We'll see what happens, although the cancelled Space Shuttle program announced in 2011 doesn't bode well.

An excellent website at www.paleofuture.com lists many predictions made from the 1870s onwards complete with fascinating diagrams. It is well worth a visit.

So what about inter-stellar and inter- and intra-galactic journeys by faster than light travel or worm holes or by other yet-to-be discovered technologies? Obviously we can't say that these things will never happen but based on the speed of progress in space travel over the past 40 years and what we know about future plans, we can probably predict that these developments are at least many hundreds or more likely thousands of years away if indeed they ever happen at all.

It's a shame really. We may be forever trapped by the Laws of Physics (and economics) to remain within the confines of our own Solar System and its nearest neighbours. There could be and probably are unknown wonders within the Universe that we will never see.

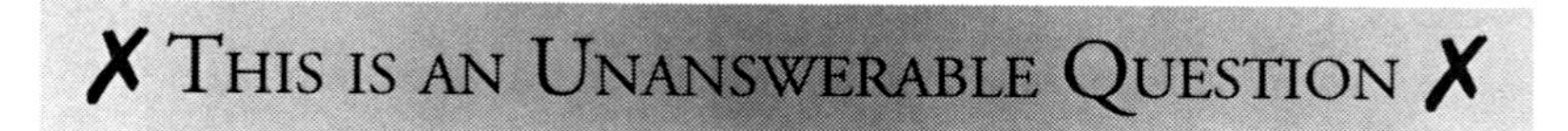
✗ THIS IS AN UNANSWERABLE QUESTION ✗

BEST GUESS ANSWER

Inter-planetary travel within our Solar System is probably do-able with Government support. Inter-stellar travel, at least to our closest stellar neighbours, may become theoretically possible with foreseeable technology but would either require many generations of astronauts to complete the journey, or a system of deep suspended animation for the entire crew, or the development of nuclear-powered constant acceleration drives.

Longer distance inter-galactic and all intra-galactic travel is currently unimaginable and will probably remain so for many thousands of years, or perhaps even for ever. But who knows what we may become capable of in the distant future?

Question 10: Is Time Travel possible?

Time, the period between two events, is usually called the 4th dimension and travelling backwards and forwards through time is a concept beloved of science fiction writers. There are many examples, including HG Wells' novel *The Time Machine* published in 1895 (Fig. 88) and filmed in 1960, as well as the various *Planet of the Apes* and *Back to the Future* films.

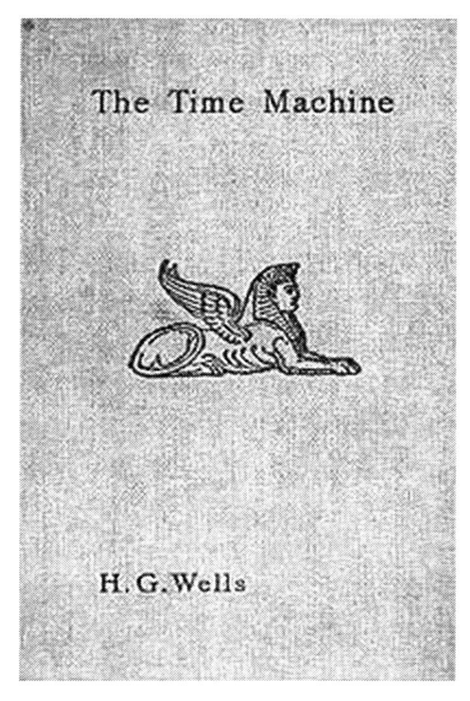

Fig. 88 The rather uninspiring cover for the 1st edition of HG Wells' The Time Machine (1895) Wikimedia Commons

But is time travel possible in real life?

Travellling back in time

It's certainly possible to *look* back in time. Since the speed of light is finite, it takes a certain amount of time to travel from its source to your eyes. When you glance at your watch to see the time, your eyes are about 30 cm from your wrist. Light takes about one billionth of a second to travel 30 cm so you are actually seeing your watch, and the time, as it was one billionth of a second ago. Whenever we look at something we are always looking back in time – it is in fact impossible to look at something and see it as it is *now.*

This doesn't make much difference in practice though. When you look at the Sun, you are seeing it as it was 8 minutes and 20 seconds ago because that is how long it takes for the Sun's light to travel to the Earth. If the Sun had suddenly disappeared just after you looked at it, you would have no knowledge of this for just over 8 minutes. It would look and feel exactly the same due to the light and heat that had already left it before it disappeared.

The light from the Andromeda galaxy takes however 2.5 million years to reach the Earth, so we're looking 2.5 million years into the past.

If we could one day develop telescopes powerful enough to see detail on stars and planets within the Andromeda galaxy, then we would be observing in real time events that took place 2.5 million years ago.

If an advanced alien on a planet 65 million light years away looked at the Earth now with a sufficiently powerful telescope, he(?) would be able to see dinosaurs roaming around (Fig. 89).

Fig. 89 View of Earth now through an alien telescope on a planet 65 million light years away

But what about *travelling* back in time rather than just looking? The so-called *Grandfather Paradox* is one of several thought experiments that make it seem that travelling backwards in time is impossible. Imagine that you were able to travel back in time and that you met and killed your own grandfather before he met your grandmother.

This would mean that your father, and therefore yourself, would never have been born. The only way to resolve this paradox is to say that travelling backwards in time is impossible.

Well, not quite the only way. You could also argue that travelling backwards in time is actually possible but that it is impossible to interfere with history. What has happened has happened and can't be changed. This doesn't seem a very satisfactory argument however. Some additional evidence, if that is the right word, against the possibility of backwards time travel is the following argument – if it was possible to travel back in time then why aren't there any time travellers here from the future?

It has been suggested that such travellers are and have been here but disguised as present day people so that we don't recognise them as from the future. That makes no sense – if you had travelled back in time you would surely wish to make contact and distribute your knowledge of what was to come.

Based on current opinion therefore, it seems that travelling back in time is not possible.

Travelling forward in time

The *perception* of time can be speeded up by hibernation or suspended animation where body temperatures and metabolic rates are reduced. Time will seem to have passed more quickly for the organism that wakes from hibernation or suspended animation, or even from an anaesthetic or from sleep.

However, this isn't really time travel since the same time would have passed whether suspended animation etc existed or not. It's only the *perception* of time that has altered.

One of the consequences of the Theory of Relativity is that time slows down as speed increases. This is known as *Time Dilation.* Although the effect is present at all speeds, it only becomes significant at speeds close to that of light.

Astronauts inside a space craft travelling at close to light speed, if that ever becomes possible, would find that everything in and on the craft showed less elapsed time than that measured by mission control and others back on Earth. In other words, time slows down for them.

They could reach several stars well within a lifetime and find that millions of years had passed when they returned to Earth. This is the basis of the *Planet of the Apes* story.

Alternatively, they could perhaps just whizz round the Earth rather than travelling anywhere. As long as they were going fast enough for long enough, they could land after say 10 of their years to find that maybe 10,000 years had passed on Earth. It's a one-way journey though, since there would be no way of travelling back to their starting date. It would be hard to find volunteers for such a trip.

Time dilation has been tested experimentally by flying atomic clocks in aircraft and showing that there is a difference in elapsed time between these travelling clocks and similar atomic clocks back on Earth. Also, atomic clocks on GPS satellites circling the Earth need corrections due to time dilation. The differences are small though, of the order of millionths of a second a day.

The Russian cosmonaut Sergei Aydeyev has spent a total of over 2 years on the Mir space station travelling at about 17,000 mph. Due to time dilation, it has been calculated that Mr Avdeyev is 0.02 seconds younger than he would have been had he not travelled in space.

Wormholes, those theoretical features of space-time that were mentioned in question 9, could perhaps provide a pathway between the present and the future. Is has to be said however, that this is currently much more in the realm of science fiction than science fact.

Whereas it seems that time travel into the past is not possible, there is at least the physical principle of time dilation that could, if the required near light speeds are ever achieved, perhaps one day enable time travel into the future. It's likely to be a one-way journey though.

If constant acceleration drives could ever be developed (see Question 9), then speeds approaching that of light could perhaps be attained. This means that a journey to Proxima Centauri, just over 4 light years away, would take about 5 years to complete.

As explained in Question 9, because of relativity, the elapsed time frame for the crew will be shorter than that for observers back on Earth. The journey time from the crew's point of view of 5 years might appear to observers on Earth as having taken 50 years.

The crew will in effect have travelled into the future. If they stay for a year and then return to Earth, they will have been away for 11 years from their point of view. On Earth however, 101 years will have passed. So, if they start their journey in 2050 and return 11 years later, it won't be 2061 but 2151.

We can end this section with a rather bizarre notion. We have seen that time travel into the past seems to be impossible but, if sufficiently fast speeds can be attained, then time travel into the future is theoretically possible. So if one day we could travel into the far distant future, then we might arrive at a time when the problems of travel into the past had been solved, and we could then go back to our time of origin. However, if the problem of backwards time travel was still insoluble, we'd have to keep going forwards until we found ourselves in a time when we could go back. Bit of a risk really.

✗ This is an Unanswerable Question ✗

BEST GUESS ANSWER

Time travel can be either into the past or into the future. Conventional time travel as depicted in science fiction stories usually involves a transport device that can carry its occupants backwards and forwards in time. Travel into the past seems to be impossible since it's difficult to get around thought experiments such as the grandfather paradox.

Additionally, if it were possible, we might have expected numerous visitors from the future but there is no evidence of such visits.

Travel into the future is allowed by the Theory of Relativity due to the effect known as 'time dilation'. As an object's speed increases, the passage of time on and in the object slows down relative to that for observers on Earth. A crew travelling at near light speed for say 10 years would find that many more years had passed upon their return to Earth.

This is not quite the same as the putative machines of science fiction since it would likely be a one-way journey with no prospect of returning to the time of origin, unless of course we arrived in a future where the problems of backwards time travel had been overcome.

Time travel as per science fiction is probably impossible. One way journeys into the future are theoretically possible although construction of such transport devices is likely to be thousands of years away if indeed they could ever be built.

Question 11: Does Astrology work?

Astrology has a long history, probably arising during the ancient Babylonian civilisations in about 1,500 BCE. Its basic tenet is that the relative positions and movements of celestial bodies, mainly stars and planets, can directly influence life on Earth and can also be used to provide information about an individual's personality and his or her present and future circumstances.

Whereas in reality the Earth moves around the Sun, seen from Earth, the Sun traces an apparent path across the sky which is known as the *ecliptic*. This is shown as an oval (**RR**) in Fig. 90.

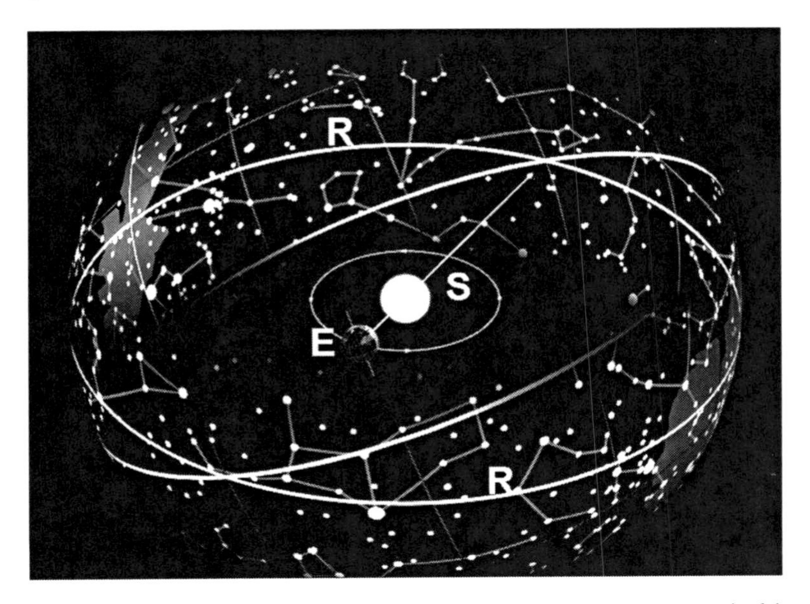

Fig. 90 Diagram showing the Earth (E) circling the Sun (S), and the apparent path of the Sun as seen from Earth during the course of 1 year (thick oval line RR). from Wikimedia Commons

This is divided into 12 regions which together make up the Zodiac. Various constellations of stars lie within each of these regions.

A constellation is a group of stars that appear to form a recognisable shape or pattern in the sky. The constellation of Leo contains some bright stars that could be joined up to form the outline of a lion; another example is Orion (Fig. 91) which can be joined up to appear like a human hunting figure wearing a belt and firing a bow and arrow. In astronomical usage however, a constellation is an area of sky and all the stars and other celestial bodies within it, so just selecting those stars that happen to form a recognisable pattern is meaningless. It's also completely arbitrary; choosing different stars within the same constellation would produce different objects.

It's also important to appreciate that although impressive patterns may be seen from the viewpoint of someone on Earth, such as the photograph of Orion in Fig. 91 (and the joined up stars to form the shape), this is a completely false view of the actual situation.

The stars that make up the pattern have no relationship with each other and are vast distances apart in all directions. Of the three named stars in the photograph, Bellatrix is closest to us at 240 light years distant (about 1,500 million million miles), whereas Rigel is furthest away at 850 light years (about 5,000 million million miles). These two stars are therefore separated by about 3,500 million million miles, not in the plane of the diagram but in depth.

Photographs and drawings made from Earth give a two-dimensional view; looked at from another viewpoint the perceived pattern would disappear completely as demonstrated in Figs. 92 and 93.

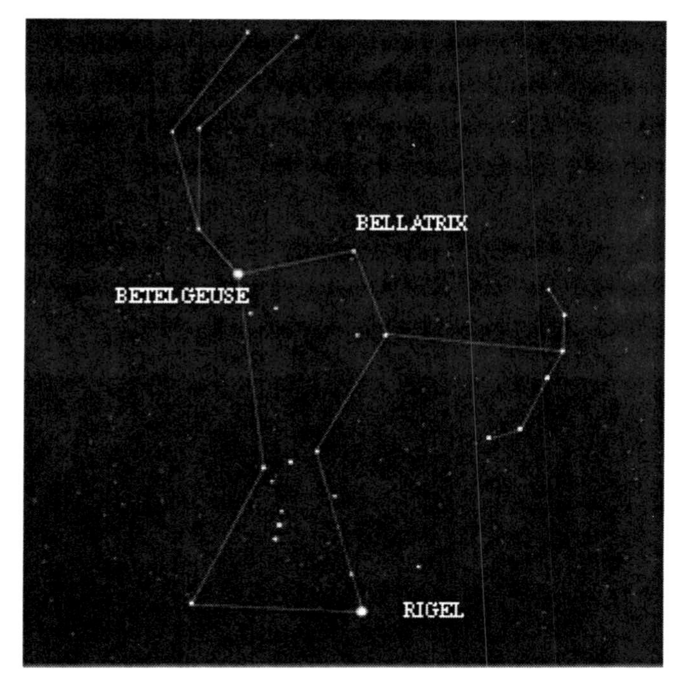

Fig. 91 Photograph of the constellation of Orion with some of the stars joined up to form the shape of a figure wearing a belt and firing a bow and arrow (Orion the Hunter). © meletver-Fotolia.com

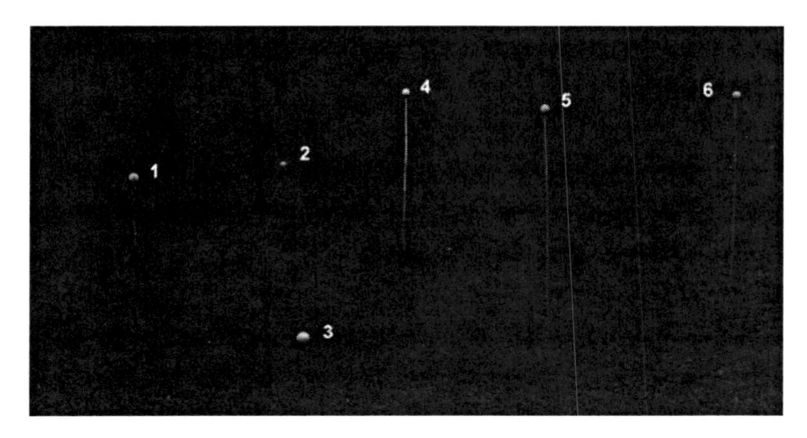

Fig. 92 'Tennis balls' constellation

Fig. 92 shows 6 tennis balls on top of bamboo poles stuck into a lawn. The photograph has been darkened to simulate a night sky, and the tennis balls are meant to represent stars. The balls appear different sizes since some are nearer the camera than others.

It's clearly possible to join the balls up to create a variety of shapes (constellations) but they have been numbered in this case so they can be identified easily in the next photograph (Fig. 93).

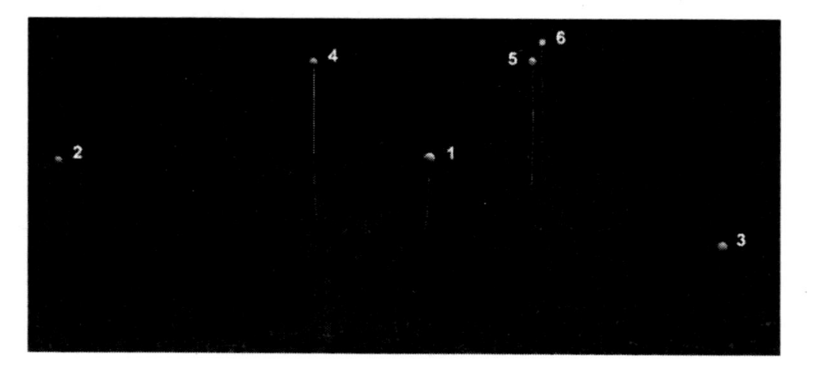

Fig. 93 'Tennis balls' constellation seen at 90° to view in Fig. 92

Fig. 93 was photographed at a 90° angle to Fig. 92. None of the poles has been moved.

It's very obvious that the entire appearance of the balls (stars) has altered. They are the same tennis balls on the same bamboo poles; only the camera position has changed.

This is exactly the situation when a constellation of stars is viewed from Earth. The apparent shape has no meaning and the constituent stars have no relationship with each other since if viewed from somewhere else the shape of the constellation would be entirely different.

Horoscopes

The 12 regions of the Zodiac, as mentioned above, together make up a period of one year, and each single region therefore consists of a period of about 30 days. Astrologers use the positions and movements of the planets within the constellations (although the planets are only *in* the constellations in the two dimensional view from Earth) as a means of predicting personal events and personal characteristics. These horoscopes (literally, *markers of the hours*) are published regularly in newspapers and magazines, and read avidly by many people.

Typical horoscopes would say things like '*be wary of offers that seem too good to be true*'; '*you may be offered a promotion at work*'; '*a lottery or premium bond win could come your way*'; '*your private life will take a turn for the better*' etc. Many of these are common occurrences and so there are good odds that at least some will happen to some people. As far as personal characteristics are concerned, astrologers would typically say that Librans, for example, symbolize a balance and always seem to be trying mentally to balance things out and get an even judgment.

However, as we have seen, the perceived shapes in the constellations, Libra the Scales in this case, are arbitrary and meaningless, so attempting to relate someone's personality to this arbitrary shape is also meaningless.

Reading horoscopes for fun is one thing but actually believing them is quite another. How could they work? The patterns are irrelevant, and the influence of far away planets and even more distant stars is hard to understand. Furthermore, before 1781, only 6 planets were known. Uranus, Neptune and Pluto had not yet been discovered, yet astrologers still claimed that they could make accurate predictions with what we now know was incomplete data.

But actually the most persuasive argument that astrology doesn't work is in fact the simplest: different readings from different sources for the same person give different results. It's a bit like asking different witnesses to a road accident the colour of the car that sped away and receiving all different answers. You wouldn't know which to believe and would have to disregard them all.

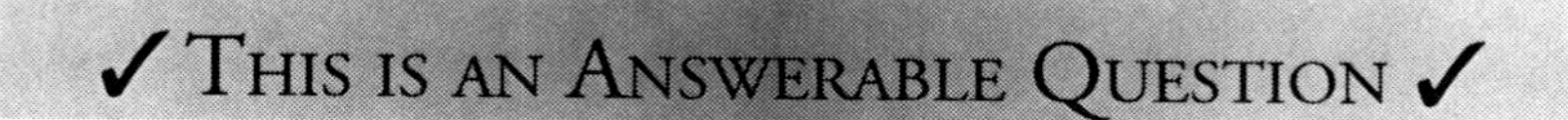
✓ This is an Answerable Question ✓

Answer

Astrology is based on several false premises and offers no plausible mechanism for its results, which are in any case often vague and ambiguous, and would apply to many people. Our answer is that it has no scientific basis and produces meaningless results and which should be used for entertainment purposes only.

Question 12: Do coincidences mean anything?

A coincidence can be defined as *an accidental occurrence of events at the same time with no apparent causal relationship.* An example might be the chance meeting of a friend in a busy shopping centre, or a sudden phone call from a relation living abroad who you had just been thinking about.

The important part of the definition is that there is no causal relationship, that is, the events happened independently by chance – one did not cause the other. If you meet a friend in a shopping centre by chance it's a coincidence; if you meet after arranging to meet, it's obviously not a coincidence.

Nevertheless, many people attach significance to coincidences believing that they were meant to be or ordained or the result of divine intervention. This is because some coincidences can indeed seem remarkable and far beyond the powers of mere chance. In most cases though this is an illusion; many things that seem unlikely are actually much more likely than people might imagine. Many people also mis-interpret or do not fully understand the power of chance.

Million to one events happen frequently as long as enough people take part. Look at the lottery where 14 million to one jackpots are won regularly because 20 or 30 million people buy tickets.

Let's have a look at a few examples of coincidences.

The birthday surprise

Ignoring leap years, how many people would there have to be in a room for at least two of them to share a birthday? Since there are 365 possible birthdays, the answer is 366 people.

But how many people would be needed for the chance of coincident birthdays to be better than 50%? The answer will surprise you – it's just *23 people*! It's true; with 23 people there is a better than evens chance that two of them will share a birthday. With 57 people, there is a 99% chance that this will happen.

The explanation requires some understanding of probability theory and is well documented in books and on the internet for those who wish to explore this further. It's a very good example of an event that seems remarkable but isn't really.

Dreams that come true

A common situation is that someone has a dream which then apparently comes true. For the dreamer, this can be a remarkable experience. Say, for example, that you have a dream about someone in a supermarket who falls over while carrying two bags full of shopping.

Then a few days later, you are in a supermarket and someone falls while carrying two large bags of shopping. What do you make of this? You will tell lots of people that you had a dream about someone in a supermarket who fell over while carrying two large bags of shopping, and it came true.

So what does this mean? Did your dream fortell the future or was it just a coincidence? Quite reasonably, you might think that it was such a remarkable coincidence that maybe it did mean something.

Most people dream about 5 times a night every night even though they may not always remember their dreams. With 60 million people in the United Kingdom, that's *300 million* dreams every night and over *100 billion* every year. That's an awful lot of dreams and some of them will bear some resemblance to actual events that haven't happened yet.

What needs to be remembered are all the many millions of dreams that *don't* come true. Someone somewhere some day will have a dream that comes true. It doesn't mean anything – it's just a statistical result from a vast number of dreams.

THE WRECK OF THE TITAN

This is the title of a book that was written in 1898 by Morgan Robertson (Fig. 94). It tells the story of a disgraced former Navy man who worked as a deckhand on a fictitious ocean liner called the *Titan* which then hit an iceberg and sank. There are many similarities between Robertson's story and the sinking of the *Titanic* 14 years later.

The *Titan* hit an iceberg in the North Atlantic and sank with the loss of over 1,000 passengers. Bearing in mind that the book was written in 1898, 14 years before the *Titanic* had its maiden voyage and even before it had been designed, the similarities between fact and fiction were indeed remarkable.

Fig. 94 The Wreck of the Titan book cover, 1898

Fig. 95 The sinking of the Titanic, 1912

The book was reprinted in 1912 after the *Titanic* disaster and some of the details in the book were changed to make the story seem even more coincidental with fact. Even so, this does not detract from the original's many similarities to the actual event.

So what does it mean? Did Morgan Robertson have the ability to fortell the future? It's unlikely. He was a merchant seaman so he knew about ships. Icebergs were a known hazard in the North Atlantic; between 1882 and 1890, 14 ships were lost due to collisions with icebergs. So he conceived a story about a disgraced naval officer who makes good with a back story about a large ocean liner that sinks after striking an iceberg in the North Atlantic.

There were indeed many similarities between the story and the real event but there were many differences as well. If it really was more than just coincidence, then we might expect all of the details to have been correct.

CHANCE MEETINGS

How many people do you know? For most of us, the answer is probably in the mid to high hundreds if we include family, friends, neighbours, acquaintances, work colleagues, and tradespeople. You can get an approximate idea as follows.

Take a reasonably common name such as Michael. Several internet sources will be able to provide data on how many people have this name in the UK; one such site is www.kgbanswers.co.uk

The number provided for Michael is 673,000 (to the nearest thousand). The population of the UK when the data were researched (2007) was about 61,000,000. This means that 1.1% of the population is called Michael.

Now work out how many Michaels you know personally. They don't have to be close friends or family; you just have to know them even if you haven't seen them for 5 years (but they do have to be alive). I could list 13 Michaels. Assuming that the percentage of Michaels in the population is the same as for my personal sample, then this means that 13 represents 1.1% of all the people I know. This calculates as close to 1200 people.

You can increase the accuracy of the estimate by repeating the exercise with some different names.

How many places do you go to and travel through during the course of a year? It's hard to give an accurate answer but if on average we are in 5 to 10 different locations each day (bus, train, shop, road, work, restaurant, cinema, club, library, airport, etc.), it would add up to about 2,500 in a year.

Chance meeting?

Let's err on the cautious side and say that most of us know about 750 people and that, during the course of a year, we spend at least some time in 2,500 different places. That amounts to 750 x 2,500 = nearly *2 million* possibilities of unexpectedly meeting someone we know per year.

Not such a chance meeting after all.

Laura Buxton's balloon

In 2001, the following story appeared in the British press.

A 10 year old girl called Laura Buxton who lives in Staffordshire released a helium balloon in her back garden. The balloon contained a note of her name and address. It landed 140 miles away in another girl's back garden. She was also 10 years old and her name was also Laura Buxton.

This certainly ranks as an amazing coincidence. Assuming that this is actually a true story and also that just one balloon with one name was released, then there is something else to consider. Laura is a popular girl's name. In 1984 it was the second most popular baby name in the UK; by 1998 its rank had dropped to 24. It's a fair estimate that in 1991, when this Laura would have been born, the rank was around 12th or 13th. Also, the surname Buxton is most common in the county of Staffordshire where Laura Buxton lived; more than 5% of all the Buxtons in the UK live in Staffordshire (data from soFeminine.co.uk). *Note: There is a lot of discussion on the internet about this event.*

This is still a remarkable event but perhaps not quite as astounding as one might first think. Also, the original story was embellished with additional data including the fact that both girls had black labradors, a guinea pig and a rabbit as pets.

This was a shame since the features of commonality had been cherry-picked. Most labradors are black; why weren't the colours of the rabbit and guinea pig mentioned? – because they weren't the same, obviously. The story was good enough with just the girls' names and should have been left at that.

THE MEISSEN PLATE

Here's a rather less remarkable story but one that happened to me. My parents had a set of Meissen porcelain which had been in the family for many years and of which they were very fond. Shortly after my mother died my wife and I were browsing round a local antique fair, as we did quite often, when we came across a stall selling a Meissen plate of the same pattern as our family set (Fig. 96).

Fig. 96 Rose pattern Meissen plate

This was the first time I had ever seen one of these plates at an antique fair – Meissen have a vast number of different designs and to find one of the same Rose pattern design as the rest of our set, and so soon after my mother had died, was quite a coincidence.

I bought the plate. It would have been very easy to think that finding this particular plate so soon after my mother died had some sort of significance – a message maybe to say that she was thinking of us and that we should keep and look after these family heirlooms.

Maybe. On the other hand, it could just have been a coincidence - *an accidental occurrence of events at the same time with no apparent causal relationship.* That seems a much more reasonable explanation to me; coincidences do happen.

TOTAL ECLIPSE OF THE SUN

The Sun has a diameter of 864,000 miles and is about 93 million miles from the Earth. The Moon has a diameter of 2,160 miles and is about 240,000 miles from the Earth. This means that the Sun is 400 times larger than the Moon and that it is on average very nearly 400 times further away. Because of this coincidence, when we look at the Sun and the Moon from Earth, both appear to be the same size. That's why we get total eclipses of the Sun when the Moon completely blocks out the Sun.

If the Earth to Moon distance had been, say, 1 million miles instead of 240,000 miles, that is, about 4 times further away, then the Moon would look about 4 times smaller. Look what happens to the appearance of the total eclipse (Figs. 97/98).

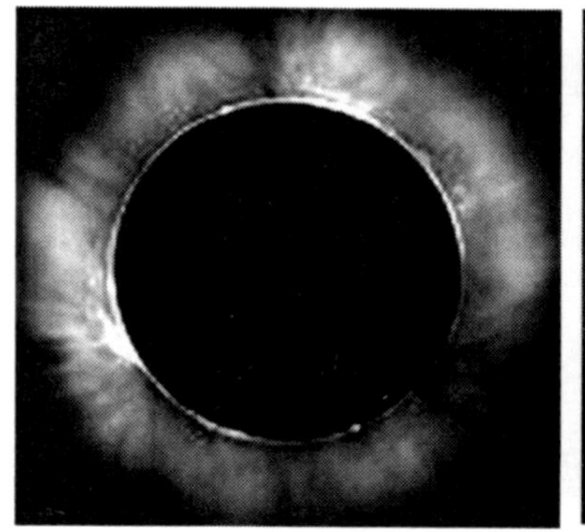

Fig. 97 Eclipse seen from Earth

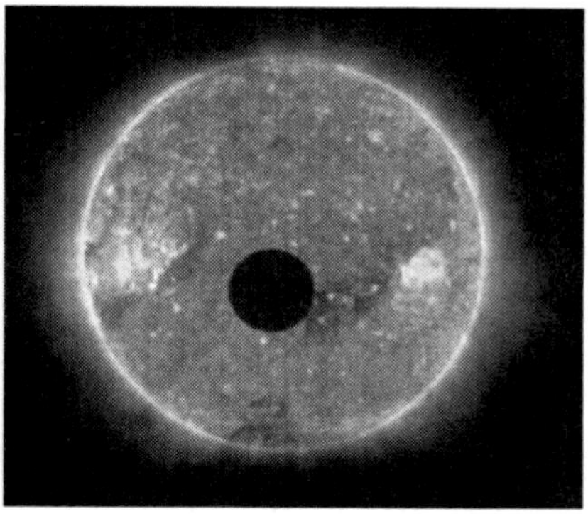

Fig. 98 Eclipse seen 1 million miles from Earth

© NASA

Fig. 97 shows a total eclipse of the Sun as seen from Earth. The Moon completely obscures the Sun because although it is 400 times smaller, the Sun is 400 times further away, so both appear the same size when viewed from Earth.

Fig. 98 shows the appearance from NASA's Stereo B spacecraft which photographed this passage of the Moon across the face of the Sun in 2007.

Stereo B is in an orbit about 1 million miles behind Earth and was 4.4 times further away from the Moon as we are on the Earth's surface. As explained above, this makes the Moon appear 4.4 times smaller when viewed from the space craft. So we don't get a total eclipse from this viewpoint because the Moon appears too small. The apparent size of the Sun does not change significantly because it is already 93 million miles away and another 1 million miles makes little difference

.

It's quite a coincidence that the apparent sizes of the Sun and the Moon as viewed from Earth are practically the same, making it possible to have a total eclipse of the Sun. No other Sun, planet, moon arrangement in the Solar System produces a total solar eclipse.

Is there any significance in this? No. It's just a coincidence. Bear in mind that the distance of the Moon from the Earth is slowly increasing by 3 to 4 centimetres per year. That doesn't sound much but eventually, in perhaps 1 billion years or so, the Moon will have moved far enough away so that its size as viewed from Earth will no longer be large enough to cover the Sun. At that time, total solar eclipses will forever be impossible.

✓ This is an Answerable Question ✓

ANSWER

Coincidences are events that are time-related but not cause-related. They happen independently and even though some may appear remarkable, there is often a statistical reason for the concordance which may not be apparent to a non-mathematician. Other seemingly remarkable coincidences which cannot be explained mathematically are far more likely to be the result of pure chance than any other more esoteric reason.

Question 13: Does prayer work?

Prayer is a religious practice that seeks to activate some form of connection to a God. It may be individual or communal and may take place in public or in private.

A prayer is likely to be offered for one of two reasons; to give thanks for good things or to ask for help for bad things. The word is thought to come from the Old French word *preier* meaning 'to ask'.

Although different religions will have their own customs and rituals and forms of prayer, the basic idea is common. The big question is, does it work?

Fig. 99 Congregants at prayer in a Christian cathedral

There's no doubt that things other than conventional medicine can make ill people better. When pharmaceutical companies test new drugs, they always run what are known as double blind randomised controlled clinical trials.

Let's say we want to test whether a new drug is effective for treating high blood pressure. We'd find say 400 people of similar ages and backgrounds with high blood pressure and divide them into two groups of 200 each. One group would be given the new drug and the other would be given a dummy pill that did not contain any drugs but looked the same. The patients would be split up randomly, and would be given a pill randomly. That means that the patients didn't know whether they had the real pill or the dummy pill (in fact, they would only be told that they were taking part in the trial of a new drug) and the doctors running the trial didn't know either. The trial runs for 3 months.

At the end of the 3 months, every patient is asked to confirm that they took their pills as required, and their blood pressure is measured and recorded. Only then are the doctors running the trial told which patients had which pills.

Here's what the results might have looked like.

TREATMENT	PATIENTS	IMPROVED	DIDN'T IMPROVE
dummy pill	200	40	160
real pill	200	150	50

There's no chance of bias here because no-one involved in the trial knew who was taking real pills or dummy pills. The results are impressive. 75% of patients taking the new pill had a reduced blood pressure at the end of the trial whereas only 20% of patients taking the dummy pill had a reduced blood pressure. So the new pill works.

But there's something else, perhaps even more remarkable. We said that *only* 20% of patients taking the dummy pill had a reduced blood pressure. But as it was a dummy pill, why should any of them have had a reduced blood pressure at all? After all, they hadn't been given anything apart from a dummy pill. It's called the *placebo effect* and is a well known medical curiosity which first came to light during World War 2 when a doctor ran out of morphine and his nurse injected the patient with salt solution pretending that it was morphine. The operation was successful with the patient not suffering any discomfort.

Placebos can be effective in many conditions and can last for a considerable time. It probably works through substances known as *endorphins* which are produced by the body under certain conditions. These are similar to morphine and other opiates, and are pain relievers and also induce a feeling of well-being.

For our purposes though, it serves to illustrate that people can be made to feel better by giving them nothing. It's the *belief* that they have been given something that probably releases endorphins and these make the body feel better. Similarly, it's undoubtedly true that many people do feel better after prayer. Their belief that they have communicated with God does the work and again, endorphins may be the mechanism.

At that level, prayer certainly does work although probably not for the reason that the worshiper believes. But what about at a more serious level? What about, for example, the Jewish prisoners in Auschwitz concentration camp praying to their God for help? Were their prayers answered? Did they feel better afterwards? No, 3 million of them died there.

Or the 800,000 Rwandan Tutsis who were killed during the genocide in 1994 and the many children who were orphaned – did they pray to their Catholic God for help? You bet they did. Where was he when his people needed them? (Fig. 100).

Fig. 100 Rwanda 1994
from Wikimedia Commons

There are of course many more examples like these where unfortunate people are overcome by such horror that the placebo effect is powerless.

And in any case, even the strongest placebo effect would be useless when confronted with a ruthless enemy rather than with an illness. All the morphine-like substances in the World can't protect you from machine guns and gas chambers.

Those who pray also do it on behalf of others, usually in the hope that this might influence the outcome of a serious illness or a major operation or a dangerous mission. It's impossible to conceive a mechanism as how this could be effective except to say that the person doing the praying may well feel better in the belief that they have done all that they could.

About 140,000 troops have been sent to Afghanistan since the war began in 2001, and 2,400 of them had been killed up to the end of May 2011. It's reasonable to assume that most of the families of these 140,000 soldiers would have been praying for their safe return, but 2,400, just under 2%, died. What do we deduce from this?

That prayer is 98% effective? Those whose sons and daughters survived would say so. But ask those whose sons and daughters died and they'd say that prayer was 0% effective.

It's far more likely that 2% represents the likelihood of being killed in such a conflict and that prayer has nothing to do with the survival rate. In any case, if prayer does work, why would God ignore the pleas of 1,500 people while protecting the remaining 68,500? There's just no sense in that.

Here's a true story. A local priest, the Reverend Neil Smith (not his real name) suffered a serious heart attack and was hospitalised for several weeks during which time he nearly died. It's fair to assume that, as a religious man, he and his family would have prayed for a good outcome. In the event, he did make a full recovery and returned to work.

I asked him whether he blamed God for his illness or thanked God for his recovery. He answered that God had better things to do than to strike him down with a heart attack for something that he may have done and that his illness was a medical matter and had nothing to do with God. However, he thanked God for his good recovery.

It's like the 33 trapped Chilean miners rescued from their mine in the Atacama desert in 2010. These are deeply religious people who clearly would have prayed for their safe return but would not have blamed God for trapping them in the first place.

However, you can't have it both ways. Either God is responsible for what happens or he isn't.

What do religious leaders have to say about why bad things happen to blameless people? Over Easter 2011, the Pope, leader of over 1 billion Catholics, agreed to answer questions from people around the World. He responded to 7 out of more than 2,000 that were submitted. The first was from a young Japanese girl who said that she was very afraid after the earthquake and tsunami which killed many of her friends and classmates, and asked why this happened.

The Pope replied that he did not know why it happened but reminded her that Jesus also suffered and that he wants to help her with his prayers and that she could be sure that God would help her as well.

Why would God help this little child if he caused her misery in the first place? A more helpful and honest answer would have been to explain that natural events happen in unpredictable ways. No-one makes them happen; they just do because that's how Nature works.

✓ This is an Answerable Question ✓

ANSWER

People who believe in a God probably feel better after prayer since their belief that they have communicated with him releases endorphins from their pituitary gland and hypothalamus into the bloodstream, and this promotes pain relief and a feeling of well-being. At this level, prayer will work for some people. At a more serious level where the threat is from a powerful and ruthless enemy (such as war, personal attack, life-threatening illness), prayer is totally ineffective in influencing the outcome.

Religious people tend to thank God for good things but do not blame him for bad things. That's like saying that we will discard all the results of an experiment that don't give us the result that we want and only consider the results that do give us the result that we want. With that attitude you could prove anything. Ask Reverend Smith, or the Chilean miners, if prayer works. Of course it does, and we're here to prove it. Ask them why they got into trouble in the first place and you won't get a straight answer, not even from as holy a person as the Pope.

Question 14:

Could we live forever?

Assuming that anyone would want to, is it possible to live forever? Medieval alchemists spent hundreds of years searching for the mythical *Elixir of Life*, a medical potion that was supposed to confer eternal youth. They never found it. Can modern science and medicine do any better?

Before we get to ways of possibly extending our life span, we need to consider how long we can expect to live without any intervention. This is known as the *life expectancy* and it varies a great deal according to the country in which you live, with the poorest countries not surprisingly having the lowest life expectancies. This is because poverty goes hand in hand with poor nutrition, poor sanitation, and poor health care.

The following table shows data from 2005 to 2010.

COUNTRY	OVERALL LIFE EXPECTANCY AT BIRTH
Japan	82.6 years
Hong Kong	82.2 years
Iceland	81.8 years
Switzerland	81.7 years
Australia	81.2 years
Spain	80.9 years
Sweden	80.9 years
Israel	80.7 years

Macau	80.7 years
France	80.7 years
UK	79.4 years
Rwanda	46.2 years
Liberia	45.7 years
Central African Republic	44.7 years
Afghanistan	43.8 years
Zimbabwe	43.5 years
Lesotho	42.6 years
Sierra Leone	42.6 years
Zambia	42.4 years
Mozambique	42.1 years
Swaziland	39.6 years

Data from United Nations records 2005 – 2010

There is a staggering 100% difference between the top and the bottom of the table showing how much still needs to be achieved to bring the bottom countries in line with the top ones.

An on-line calculator at www.calculator.livingto100.com calculates your personal life expectancy based on your answers to a number of lifestyle and medical questions.

Life expectancies have been increasing steadily over the years as shown in the graph (Fig. 101), from an average of about 74 in 1980 to an average of about 80 in 2007.

Recent government figures show that 17% of the UK population can expect to live to 100. But what if we want to do better than this?

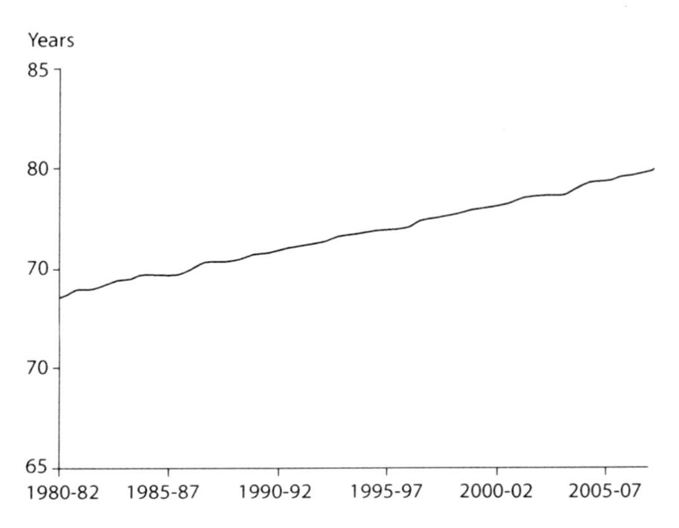

Fig. 101 Graph showing the steady increase in average Life Expectancy in the UK from 1980 to 2007. Data from Office of National Statistics

What if we want to live for 200 years or even longer? Is this possible? There are accounts of bacterial and other spores being extracted and revived from mineral deposits and from amber after hundreds of millions of years. Such organisms are able to survive in a state of deep suspended animation for extraordinary amounts of time. The oldest living things ever recorded are some bacteria *Bacillus permians* which were revived after being found embedded in salt crystals in New Mexico in 2000. They were dated as being 250 million years old.

Some pine and spruce trees have been found to be several thousands of years old. Animals struggle to reach 200 years and the oldest human to have lived, with proper documents to prove it, was Jeanne Calment, a French lady who died in 1997 aged 122 years and 164 days. Can this be improved upon? Could we live, if not forever, for hundreds or even thousands of years?

More people could certainly live longer if they adopted a healthier life style but this is unlikely to result in significant increases in life span.

There is a very rare genetic disease known as *progeria* which affects about 1 in 8 million people. Those born with this condition age much more quickly than normal people and usually die in their teens due to diseases of old age, such as hardening of the arteries and heart failure. As doctors and scientists learn more about how this disease speeds up the ageing process, they may also find ways of slowing it down.

Research work on *telomeres* may prove rewarding. Telomeres are the end bits of chromosomes rather like the metal tips of shoelaces (Fig. 102). Both serve similar functions in that they prevent the ends from fraying. Telomeres protect the ends of chromosomes during the duplication process, and become shortened after each cell division.

Eventually they become so short that they can no longer function properly, and the cell cannot duplicate and dies. This is possibly either a cause or a consequence of ageing.

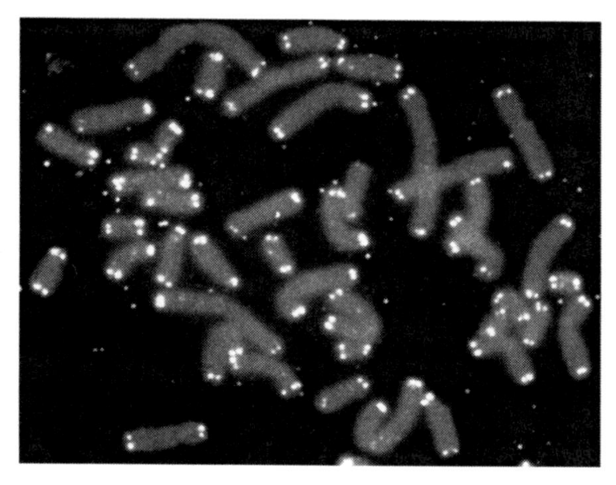

Fig. 102 Human chromosomes showing telomeres as white spots. Wikimedia Commons

There is a limit as to the number of times a cell can divide before it stops, presumably because the telomeres have become too short. For human cells, this is between about 60 and 80 times. Telomeres can be lengthened by an enzyme, *telomerase*, and it has been suggested that increasing the length of the telomeres might enable cells to divide for longer and the ageing process would be slowed down. Interestingly, telomeres from patients with progeria are shorter than those from normal people of the same age.

In 2011, a Spanish company *Life Length* announced the availability of telomere length measurements and a calculated estimate of life expectancy. This won't of course tell you if you will be run over by a bus.

If it turns out that a shortened telomere length really is a cause of ageing (this isn't proven yet – shorter telomeres could be a *result* of ageing rather than a *cause*) then telomerase may provide a mechanism whereby the ageing process really could be slowed down. This could be a modern day Elixir of Life.

It's likely to be a long way off though since even if a way of lengthening telomeres in humans can be found, there would need to be rigorous studies to show that it is safe to do so. One potential risk would be cancer. Cancer cells divide uncontrollably and have longer telomeres than normal cells.

How much could human life spans be increased if one day the telomere method can be shown to be safe and to work? It's hard to say but it could conceivably be several hundreds of years.

We'll have to leave aside the difficult questions of how to house and care for and fund the rapidly expanding and aging population that would result from this when we already seem to have too many people in the World.

Simpler and safer techniques may already exist. It's been known for many years that laboratory animals live longer on a calorie restricted diet although evidence that this also works in humans is lacking.

However, even if it were true that the human life span could be extended by say 10 or 20 years by eating a calorie restricted diet for 50 years, it's doubtful that anyone would take this up. Most people enjoy their food too much to give it up for 50 years on a promise of a few more years of life.

In 2010, it was reported that the lifespan of mice could be extended by about 12% when they were fed with water containing some essential amino acids. They were also stronger and had more stamina than the control group. This has yet to be tried on humans but if proven, could be a cheap and acceptable way of increasing life span and health in old age.

Finally, mention should be made of *cryopreservation* although this isn't really the same as increasing life span. Cryopreservation, which has some popularity in the USA, involves freezing a body just after death in the hope that it can be revived at a later date when there is a treatment for the condition that caused its death. In some cases, just the head of the body is preserved. One company offering such facilities is the Cryonics Institute in Michigan, USA (Fig. 103).

Fig. 103 The Cryonics Institute, Michigan, USA

For lifetime membership there is a one-time fee of $1,250 followed by a fee of $28,000 for the actual cryopreservation. In May 2011, there were 920 paid-up members with 103 frozen humans and 76 frozen pets. The founder of the Cryonics Institute, Mr Robert Ettinger, died on 23 July 2011. He was returned as a customer the next day.

Most scientists and doctors consider this to be a fanciful aspiration. One of the many difficulties is that as cells freeze they form crystals of ice, and these will destroy the structure of the cells making them unviable.

Even if ice crystal formation could be prevented by use of anti-freeze agents or by special cooling processes, there is absolutely no evidence that an organ as large as a human brain could be resuscitated successfully, never mind an entire human body. Even so, as mentioned above, a few very wealthy people have signed up for this process.

As soon as possible after death, a member patient is infused with a substance intended to prevent ice formation, cooled to a temperature of -196° C where physical decay practically stops, and then maintained in liquid nitrogen at this temperature indefinitely. If and when future medical technology allows, the patient hopes to be healed, rejuvenated, revived, and awakened to a greatly extended life in youthful good health, free from disease or the ageing process. That's what it says in the brochure. So far, it's never been done.

✓ This is an Answerable Question ✓

ANSWER

No human has lived longer than 122 years, at least not with proper documentation to prove it. A healthy lifestyle will help more people to live longer but is unlikely to increase significantly the current life span. New technology would be needed for that to happen. Since ageing is a process, it should theoretically be possible to interfere with it and slow it down.

Research on telomeres may be our best prospect so far and it is feasible that life spans could be increased several fold once the process is completely understood, known to be safe, and becomes controllable.

It currently takes 10 to 12 years for a drug company to progress from a possible new medicine in a test tube to a tablet on the market. Life span extending medication is theoretically feasible and could lead to human lifetimes perhaps into the hundreds of years. There are huge problems however, not the least being the safety of such interference in biochemical processes.

It would take several decades to be sure that, for example, a dog that was made to live for 50 years (the record for the World's oldest dog, who died in 2010, is currently 21 years) remained healthy and without side effects. Human testing could not start until researchers and safety committees were satisfied that extensive animal testing had shown the treatment to be safe.

Such testing would itself take many decades to complete, so it's unreasonable to expect telomere-based or indeed any other anti-ageing treatment for increased life span to become available for at least 100 years if indeed it is ever permitted.

However, the social and economic problems created by a population containing 500 year-old multi-great grandfathers and mothers are likely to be insurmountable. Of course, not everyone would wish to live that long but some might. Perhaps there would different strengths of tablets offering differing life spans. 100mg for every 100 years maybe.

Calorie restriction offers a safe and cost-free option but is unlikely to become a popular remedy. Food supplement cocktails, such as the essential amino acids which apparently worked with mice, would also be a safe and low cost option, and one which would be eagerly taken up by many people. Further research data are urgently needed.

Even if both of these latter options turn out to be effective, the increase in human life span is likely to be of the order of 10 to 20 years or so. While still significant, this is a far way off from the idea of living for an extra hundred years or more.

Our answer is that life span enhancing procedures for a modest (tens of years) increase in life span are feasible. More extensive increases in the 100 or more year range may become theoretically possible but are unlikely to become available for at least 100 years if they are allowed at all. Living *forever* is almost certainly impossible. Cryopreservation of brains and whole bodies after death is likely to be a waste of money and resources.

Question 15:

What happens when we die?

This, and the following question about a Creator, really are perhaps the biggest questions of all. They have occupied the minds of scientists, philosophers and religious leaders for thousands of years yet they remain a mystery.

The reason for our ignorance about death is simple; no-one has come back to tell us the answer.

Basically, there are just two possibilities. Either we die and our existence ends at all levels, or we die and part of us moves on to a different level of existence which has been called the *afterlife*.

When we fry a piece of cod to have with our mushy peas and chips, we don't think about the fish as a living creature any more. And why should we? Its cells have been without oxygen for several days and we have just heated it in oil to nearly 200°C. No Earth-bound life could survive that. Similarly, when a tree is chopped down, or when a colony of bacteria is exposed to antibiotics, their life ends.

Common to these situations is that all biochemical reactions have ceased and as a consequence the vital processes necessary for life cannot continue. That's what we call *death*. It is the same outcome whatever the life form – an ant or an antelope; a man or a mango. No-one would argue with this. Death is death. The difficulty arises when we ask – what happens next?

There are two options.

NOTHING

Well, not quite nothing because the cells of the now dead creature will begin to decompose leaving any hard components such as shell and bone to survive for many years, depending on the local conditions. But that's it. Parts of the body may remain for some time but for the organism itself, its existence, at any level, is at an end.

SOMETHING

Many people, possibly the majority, believe that *something* follows after death. This is one reason why it has been possible to persuade so many people to become martyrs for their cause; they have a fervent belief that they will enter Paradise after death. A general term for this 'something' is the *afterlife*.

The afterlife refers to the continuation of some part of a person's consciousness, sometimes referred to as the soul, after his or her physical death. The ancient Egyptians were perhaps the greatest exponents of this concept, constructing enormous pyramids and burial chambers to house the mummified bodies of their pharaohs (Fig. 104). Many precious objects were interred with the body for use in the afterlife.

They also created a large series of papyri containing details of rituals and spells which has become known as *The Book of the Dead*. It was believed that these would help them on their journey to the afterlife.

Fig. 104 The great pyramid at Giza, built about 2500 BCE

Most religions encompass a belief in an afterlife. And what an engaging concept it is, existing forever in a trouble-free environment surrounded by long lost loved ones. But is it true?

Is there really a part of the body, or soul, that leaves when you die for a new existence elsewhere? Apart from wishful thinking, is there any evidence for this idea?

In 1907, Dr Duncan MacDougall, an American physician, weighed 6 patients in an old age home who were dying from tuberculosis. He placed the patients, still in their beds, on an industrial scale and weighed them before and after they died.

On average, he found that there was a weight loss of 21 grams after death, and he interpreted this as support for his idea that the soul had a weight, and that it was 21 grams (Fig. 105).

SOUL HAS WEIGHT, PHYSICIAN THINKS

Dr. Macdougall of Haverhill Tells of Experiments at Death.

LOSS TO BODY RECORDED

Scales Showed an Ounce Gone in One Case, He Says—Four Other Doctors Present.

Special to The New York Times.

Fig. 105 Report in the New York Times, 10 March 1907. © New York Times

To suggest that an industrial scale set up in an open room in 1907 would be sensitive enough to detect a weight change of 21 grams in a combined patient and metal bed weight of perhaps 200 kilograms (about one part in 10,000) is just fanciful, and Dr MacDougall's results are considered as meaningless. It's become a piece of folklore however, and even spawned a 2003 film entitled *21 grams.*

To be sure, as Hamlet said, *There are more things in heaven and earth, Horatio, than are dreamt of in your philosophy.* However, that doesn't mean that we have to accept everything; we should ask for some evidence at least.

Does the soul live on? Does it come back reincarnated as someone or something else? One day, somewhere in space, will we come across the souls of every human being who has ever lived (estimated at around 100 billion people)? It would be nice, maybe, but there's just no evidence to support the idea. It's a belief strongly held by many people but that doesn't make it true.

Harry Houdini, the escape artist and magician, died in 1926, and before his death, he made a pact with his wife Bess that if he could, he would return and make contact with her. They created a coded message that was only known to the two of them so that Bess could be sure that any contact really was from her husband.

Every Halloween since 1927, a séance has been held to see if Houdini would try to contact his wife (Fig. 106).

Fig. 106 The final Houdini séance with Bess (1936)
© houdinitribute.com

After 10 séances in 10 years, no message had been received, and Bess gave up trying to contact her husband.

The séances continued in Bess's absence however, but no contact was ever established. If Houdini couldn't come back, what chance have the rest of us?

In spite of the lack of any evidence, the afterlife remains a firmly held belief by vast numbers of people, probably because it's such a comforting thought. Who wouldn't want it to be true? Wanting something to be true though doesn't make it so, nor does its mention in numerous ancient religious books and manuscripts.

An excellent series of articles summarising various religions' views of the afterlife can be found on the website www.library.thinkquest.org/16665/afterlifeframe.htm

Finally, we should consider the concept of reincarnation (literally *entering the body again)* which dates back several thousands of years. Today it is accepted by many religions and significant percentages of people are believers.

Basically, reincarnation is the belief that when a person dies, part of the person is reborn in another body. This new body therefore would have a previous life and will also, at some future time, live again in another body.

It may be a consoling thought but there is absolutely no credible scientific evidence in favour of reincarnation. That doesn't stop believers from insisting that it is true however.

For example, they may cite examples of people who have memories of past events that they could not have known about and that were later verified (but why couldn't they have known about them? Books, newspapers, TV programmes, the internet – there are so many sources of information. Who can be sure that they didn't hear about or read about it somewhere?)

✓ This is an Answerable Question ✓

ANSWER

Vast numbers of people believe in some form of afterlife even though there is no credible evidence in support of such a concept. *Belief* in this context does not require evidence to back it up – people believe not because they have been shown evidence in favour but because they have been brought up and perhaps educated to believe. Evidence, or the lack of it, is irrelevant.

However, our rules of engagement in this book state that we must base our conclusions on evidence. So even though it is a comforting thought for many people, and many ancient religious books make mention of it, until we have good evidence in favour we have to say that there is no such thing as an afterlife or reincarnation. When you're dead, you're dead, and that's it.

A rebuttal to this conclusion could be that there *is* an afterlife but it is somehow beyond our event horizon (see Question 2) so that we can never know of its existence. Maybe, but since we are basing our answers on current scientific knowledge rather than speculation we have to stay with the answer as given.

QUESTION 16:

IS THERE A CREATOR?

For many people, this is probably the biggest question of all and entire books have been written on this one subject. Before we try and give our Best Guess answer, we need to agree on our definition of a Creator. To most people, *Creator* is synonymous with *God* since those who believe in God also believe that his first task was to create the Universe.

Most people would say that God is the supernatural creator of the Universe and Life and is an all-knowing, all-powerful, and perfect entity. We shall use this definition which probably covers the views of most people with these beliefs.

Let's consider how the concept of worshiping gods might have arisen. Obviously we can't be sure about this but imagine for a moment that you are a Neanderthal Man walking around somewhere in what is now present-day Europe (Fig. 107).

Fig. 107 Artist's impression of Neanderthal Man. © Hunta-Fotolia.com

It's 100,000 BCE and you don't know much about the world around you. You know how to make basic tools, how to kill and eat animals, how to shelter in a cave, and you know that it gets light and dark, and hot and cold.

You've also noticed a big and bright object in the sky that seems to appear, move across, and then disappear. You've connected its presence with warmth and light and its disappearance with cold and darkness, and you've become quite fascinated by it.

Then one day, while gazing up at it, something simultaneously amazing and terrifying happens. It starts to disappear when it shouldn't. It's like something is taking bites out of it. It starts to get very cold and very dark. You're afraid. Your heart beats faster and your breathing gets deeper. You start to sweat and panic. Then, after a few minutes, everything goes back to normal again.

We know this as an eclipse of the Sun. All that has happened is that the Moon has passed between the Earth and the Sun, blocking it out.

There were plenty of other things to frighten and worry our ancestor - thunderstorms with bolts of lightning, erupting volcanoes, tidal waves, hurricanes, etc. What would prehistoric Man have made of these events? Maybe he would have thought that they were a punishment for a bad deed, or a warning.

These events would have been experienced time and again during Man's evolution, and at the time of the first civilisation around 6,000 BCE, usually ascribed to Mesopotamia in present-day northern Iraq, they were considered to have some purpose.

The Mesopotamians worshipped many gods, and it's easy to see how the population might have derived peace and comfort from such worship when confronted with a major and unexplicable disruption to their normal day. The transient nature of astronomical and climactic events meant that things eventually returned to normal. For them, this was a persuasive indicator of the power of their prayers. Of course, had they done the control experiment, where next time there were no prayers, the future of the god concept might have taken a different turn.

Ancient civilisations such as the Mesopotamians, the ancient Egyptians, and many others, tended to worship many gods. These were individually responsible for separate things such as love, death, the Sun, the Moon, the weather etc., and were created as physical entities and/or paintings.

The concept of just one super-god responsible for everything probably arose during the later parts of the ancient Egyptian civilisation leading eventually to the widespread development of mono-theistic religions.

It's probably a reasonable assumption to say that the God concept arose as a security blanket for ancient civilisations when confronted with unknown and worrying situations.

Religions

A religion can be defined as a specific system of beliefs and ethics. A bit of research on the Internet will tell you that there are over 4,000 different religions in the World.

This figure would include all the many variations of the main religions together with a large number of other groups who have their own specific beliefs and rules.

Let's start by having a brief look at the major religions of the World. We'll call them *Beliefs* since this will then encompass atheism. This isn't to disregard the many other beliefs but to be practical – this is just one chapter.

Major Beliefs by approximate Number of Followers

BELIEF (in all its forms)	NUMBER of FOLLOWERS
Christianity	2,000,000,000
Islam	1,500,000,000
Hinduism	900,000,000
Atheism	800,000,000
Chinese traditional (folk)	400,000,000
Buddhism	380,000,000
Sikhism	25,000,000
Judaism	15,000,000
TOTAL	6,020,000,000

This total represents over 85% of the World's population (US Census Bureau current estimate for World's population = 7,000,000,000).

The major beliefs of the World are mono-theistic but differ in their concept of God. Here is a brief summary.

Christianity

Christians believe that God is a single being who exists as a fusion of three beings – the Father, the Son, and the Holy Spirit. This is the doctrine of the Trinity. God is seen as the creator of the Universe although his physical form is not defined.

Islam

Islam is based on revelations received by the Prophet Muhammad which were later recorded in the Qur'an (Koran). Followers of Islam (Muslims) believe that God (Allah) is the all-powerful **creator of the Universe**. Islam teaches that God is perfect, eternal, self-sufficient, and omnipotent, and that he does not resemble any of his creations.

He is described as being aware of everything that happens including private thoughts. His actual nature remains ultimately **unknowable**.

Hinduism

Hinduism recognises Brahman as the spiritual underpinning to everything and everyone. Most Hindus relate to this spiritual dimension through a variety of personalised forms. Brahman is not a personality so cannot be thought of as the creator of the Universe. Hinduism by-passes the idea of God as a creator and considers the creation as having been spontaneous and of unknown methodology.

Atheism

Atheism is defined by what its followers do not believe as opposed to other beliefs which are defined by what they do believe. What atheists have in common is that they do not believe in God and therefore do not believe that the Universe was created by a god. As a group there will therefore be a variety of opinions as to the method of creation although none will invoke the involvement of a supreme being or entity.

Chinese traditional (folk)

Chinese folk religion is based on the worship of hundreds of cultural heroes, deities, and supernatural beings that vary according to geographical and local conditions. It is not organised in institutions and has no clergy or formal rituals although it does have its own temples.

As far as the creation of the Universe is concerned, the belief is that in the beginning there was nothing except a formless chaos. This coalesced into a black egg in which heaven and earth were mixed, and inside this egg was an entity known as Pangu, usually depicted as a primitive hairy giant. Pangu broke out of the egg and with the contents created heaven and earth.

Buddhism

Buddhism is based upon the teachings of an Indian prince, Gautama Buddha. These were more than a set of instructions since they also implied a way of life leading to freedom from confusion and suffering. Buddha taught that there is no creator God and that speculation about the origin of the Universe was a pointless and non-rewarding pursuit.

Sikhism

The Sikh view is that the Universe and everything in it is the work of a creator who is omnipresent and has infinite qualities. Since the creator is present everywhere, he/it therefore exists within the person as well. Sikhs tend not to use the term God, since this implies a separation from them whereas the concept is one where the creator is everywhere including within the person.

A full understanding of the creator is held to be beyond human comprehension. Sikhs do not concern themselves with a mental image of the creator's physical form.

Judaism

Judaism believes in one God who must exist since without his intervention, the Universe could not have formed. God is neither matter nor spirit but is the creator of both although he himself is neither. There is God himself, who is unknowable, and the revealed aspect of God which interacts with Mankind. It is forbidden to represent or visualise God in a physical form.

Following is a summary table.

Belief	Creation of the Universe
Christianity	creator God
Islam	creator God (Allah)
Hinduism	spontaneous
Atheism	unknown but not by creator God
Chinese traditional	creator God (Pangu)
Buddhism	not considered; no creator God
Sikhism	creator: don't use the term God
Judaism	creator God

Christianity, Islam, Chinese traditional beliefs, Sikhism and Judaism, with almost 4 billion followers between them representing almost 60% of the World's population, all believe that the Universe was the work of a creator whether or not he is addressed by the term God.

The remainder either believe that the Universe always existed, or that it arose by some other as yet unknown way, or they don't know and don't need to know.

Different beliefs differ in their customs and traditions and modes of worship. However, the three main ones, currently representing almost 60% of the World's population, all believe in a creator God who responds to prayer and who is responsible for life and the Universe.

Let us now examine the fundamental tasks that believers demand of their Creator.

The creation of the Universe
The creation of Life
To act upon prayers

The creation of the Universe

We've seen in Question 1 that the Universe either arose about 13.7 billion years ago as a consequence of the laws of physics due to the conditions that prevailed at the time, or it has always existed.

To be fair, we don't know which of these is correct, or what the conditions were 13.7 billion years ago, but since the existing Universe obeys the laws of physics, it's reasonable to assume that the pre-existing Universe did as well even though they may have been different laws. Under certain conditions, possibly very unusual and very rare conditions, the Universe just had to appear. It's possible that the experiments currently being conducted by CERN at the Large Hadron Collider may give us some intriguing new information about this but for now we just don't know the physics of how it happened. What we do know is that it did happen. And it only had to happen once.

For ease of communication and discussion, most processes are given names. The name currently given by cosmologists to the process that resulted in the origin of the Universe is *The Big Bang* because this best describes what is believed to have occurred.

However, if the Large Hadron Collider experiments do come up with some definitive results, we may be calling the origin of the Universe something like *The Boson Hadron anti-Spin Interaction.* It really doesn't matter – it's just a name. If you want to call it God, then that's fine as well. Remember though that it's just a name given to a process obeying the Laws of Physics. It's not an entity in the sense that it knows who you are and what you are and what you do and what you want. There's absolutely no evidence that it will be aware of and respond to your prayers.

The creation of Life

Question 3 considered how life might have formed on Earth. Plausible mechanisms have been suggested which only require the right ingredients and the right atmospheric conditions. When these all come together, it's possible that life could form on its own. It may well have been a very unlikely event but that doesn't matter since, like the Universe, it only had to happen once.

Once again, we're talking about a process which is generally referred to as the *Origin of Life*. But that's just a descriptive name, like the *Big Bang*. If you want to call this process *God*, then that's fine. However, doing so doesn't change the fact that it happened on its own according to the Laws of Physics and Chemistry.

To act upon prayers

Those who pray may feel better in themselves afterwards but this is likely to be akin to the placebo effect which is a well-known phenomenon in medicine. Prayers designed to help those in severe distress do not provide to everyone the help that is sought after. If a God could indeed make a response, why would he be selective?

Some difficult questions

To say that something has been *designed* rather than created naturally implies that some special expertise has been employed during construction to ensure that the article fulfils its function in the best way possible, given contemporary technology and resources.

If the Universe and everything in it was indeed designed by a perfect Creator, then we can ask some questions about the details of the design. Here are a few sample questions:

- Why make the speed of light unsurpassable and the Universe so big so that travelling inter-stellar and intra-galactic distances becomes impossible?
- Why design stars, including our Sun, so that they will eventually explode as super novae and all life in their solar systems will die?
- Why have asteroids and comets which can, and have, destroyed many species of life with mass extinction events due to collisions?
- Why design a genetic process that doesn't fail safe and can subject its carriers to the misery of incurable genetic diseases?
- Why design humans with personalities that prevent all people from living in harmony?

The classic answer to such questions is that we can't and musn't question God's motives. But that not an answer; it's a cop out because there is no answer.

✓ This is an Answerable Question ✓

ANSWER

Everything that is ascribed to the power of a Creator by those who believe in him is explainable to a greater or lesser degree by science. The creation of the Universe and the creation of Life are just processes obeying the Laws of Physics and Chemistry. We may not fully understand them but, like the card trick described in Question 1, that doesn't mean that there isn't a proper explanation. We just haven't worked it out yet. So, were the Universe and Life the work of a Creator God? Only if you wish to use that name for the processes by which these events happened.

But they are just processes. There's absolutely no justification for elevating them to supernatural entity status and imbuing them with absolute power over everything that has happened and will happen. And there's certainly no justification for assuming that they will be able to receive and respond to prayer.

What's the sense in designing a Universe so large and imposing an absolute maximum speed limit within it so that most inter-stellar and all intra-galactic travel are forever forbidden? Why design the Sun so that it will eventually die along with all life on Earth?

Our answer therefore is that God, as an all-powerful all-knowing super-being responsible for the creation of the Universe and Life and who receives and responds to prayer, does not exist.

It is likely that the Universe and everything in it will one day be completely explainable by contemporary scientific principles.

Question 17: How and when will the World end?

The End of the World isn't something that most people think about since it is a depressing thought. However, one occasionally sees sandwich men displaying this message (Fig. 108).

Fig. 108 Well, in a few billion years perhaps (www.amigos-museu-sbras.org)

So what can we say about this? There have been numerous prophecies regarding the end of the World by religious groups and cults as well as by historical figures such as Nostradamus in the 16th century. An interesting list of possible cataclysmic events can be found at www.aquiziam.com/end_of_the_world_predictions.html

So far, obviously, none of these prophecies have come true since we're all still here. We'll have to wait and see what happens when some of the future dates are reached although the vast majority of the population do not take these prophecies seriously.

We can however approach this topic rather more scientifically. The World, that is the Earth and the rest of the Universe, had a beginning as discussed in the first 4 questions of Part Two so it's not unreasonable to suppose that they will also have an end. What form might this take and when might it happen?

Let's first list the possibilities. These divide into man-made scenarios and natural scenarios.

MAN-MADE SCENARIOS

Global biological, chemical or nuclear wars

It is conceivable that a global-scale war involving the use of some or all of the above weapons could destroy all life on Earth. With current sophisticated monitoring facilities however, it seems unlikely that one nation would be able to mount a surprise attack that would devastate the entire planet. In any case, such a result would hardly be of benefit to the aggressors. So while it obviously remains a possibility, the prospect of Man destroying all life on Earth through an act of war, or perhaps through a genuine mistake, seems unlikely. Our biggest threats actually come from a variety of possible natural occurrences.

NATURAL SCENARIOS

Astronomical impact

There is a large variety of objects moving around the Solar System apart from the obvious Sun and its planets and moons.

These range from minute dust particles to gigantic lumps of rock. Space explorations over the past decade have enormously increased our knowledge about these other objects but our concern here is with those that could conceivably collide with the Earth, that is, asteroids, comets, and meteors.

Asteroids are small (at least, relative to planets) rocky bodies that orbit the Sun between the orbits of Mars and Jupiter, as shown in Fig. 109. There are many millions of them. Ceres, the largest with a diameter of about 590 miles, was discovered in 1801, and it will be visited by NASA's Dawn space probe in 2015. The others are smaller, ranging in size from tens of miles downwards.

The larger asteroids, such as Ceres, are spherical but the smaller ones are irregular in shape since their gravity was insufficient to force them into a sphere when they were formed.

For example, asteroid 243 Ida (Fig. 110) has an elongated shape and is about 30 miles long. It is basically a gigantic lump of rock and is unusual in having a moon.

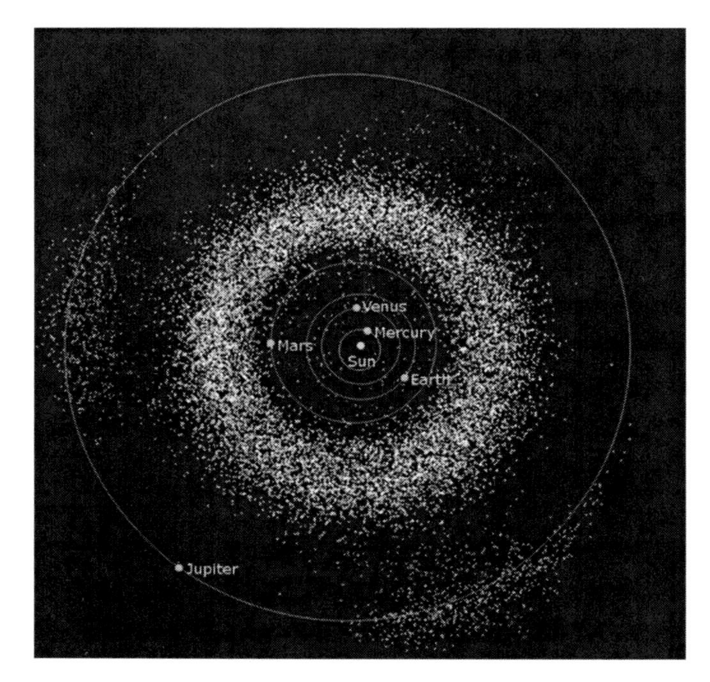

Fig. 109 The asteroid belt between Mars and Jupiter

Fig. 110 Asteroid 243 Ida and its moon Dactyl. © NASA

There is a second band of asteroids that exists outside the orbit of Neptune which is known as the Kuiper belt. Discovered in 1992, the Kuiper belt is similar to the asteroid belt only much larger – at least 20 times as wide. Pluto, re-classified as a dwarf planet in 2006, is a member of the Kuiper belt.

Comets are lumps of rock and ice that can be tens of miles wide. They are characterised by their long tails composed of ice and gas vapours produced as they pass close to the Sun. These tails can be enormous, up to 50 million miles long although the solid part of the comet, the nucleus, may be only a few miles long.

There are about 4,000 known comets but undoubtedly there are thousands if not millions of others that remain to be discovered.
The most famous comet is that named after Edmond Halley, the British astronomer who first worked out, in 1705, that this comet would return to be seen from Earth approximately every 75 years. The most recent appearance was in 1986. Fig. 111 shows it during its previous visitation in 1910.

Fig. 111 Halley's comet 1910. Yerkes Observatory

Halley's comet, although it wasn't called that then, was depicted on the Bayeux tapestry (Fig. 112).

Fig. 112 Bayeux tapestry showing spectators pointing at Halley's comet (top right) during its appearance in March 1066

Meteors A meteor (also called a shooting star) is actually the streak of light that occurs when an object, a *meteoroid*, enters the Earth's atmosphere (Fig. 113). Most are about the size of a beach pebble and will usually disintegrate during their passage through the atmosphere. Any that do land intact are known as *meteorites.*

Asteroids, comets, and meteoroids can and do collide with each other, and with planets and moons. The familiar cratered surface of our Moon is due to multiple collisions with such objects in the past. In 1994, Jupiter suffered an impact from the Shoemaker-Levy 9 comet, which was observed by NASA spacecraft cameras. Even asteroids have been hit, as is clearly seen in the cratered surface of 234 Ida (Fig. 110).

The Earth has also had its share of impacts (Fig. 114).

Fig. 113 Leonid meteor and its trail, 2009. From Navicore @ Wikimedia Commons

Fig. 114 Meteor crater, Arizona, USA. From Wikimedia Commons

This is Meteor Crater in Arizona , and a fine example of the results of an impact. The crater is 600 feet deep and ¾ mile wide, and was formed about 50,000 years ago by a meteorite estimated to be about 100 feet wide and weighing over ¼ million tons. It is currently privately owned.

One of the largest impact craters on Earth is the Chicxulub crater buried underneath the Yucatan Peninsula in Mexico. It is more than 100 miles wide and was formed 65 million years ago. Although not proven, many scientists believe that this collision caused a series of mass extinctions of plants and animals, including the dinosaurs.

A remarkable recent discovery is the Kamil crater in the Egyptian desert. This is a 45 metre wide and 16 metre deep impact crater that was discovered in 2008 by an Italian scientist Vincenzo de Michele while looking through Google Earth satellite images.

Ground-based studies of the crater in 2010 revealed that it was caused by an impact with an iron meteorite weighing about 5 to 10 tons with the impact occurring approximately 5,000 years ago. The meteorite is estimated to have been about one to two metres wide.

If an object, whether an asteroid, meteoroid, or comet, is in an orbit that brings it close to the Earth, then it is known as a Near Earth Object. It is comforting to know that several nations are undertaking a continuous monitoring of these objects especially those of a size sufficient to cause major damage on impact.

As might be expected, there are far more smaller objects that collide with the Earth than larger ones. Here is a table showing the approximate relationship between size (in metres) and frequency of impact.

Diameter	Frequency
< 10m	every year
10 - 50m	every 5 years
100m	every 1,000 years
1 – 2km	every ½ million years
15km	every 50+ million years

approximate frequency of asteroid impact depending on size (data taken from www.risk-ed.org)

Smaller objects would be likely to vapourise in the atmosphere and cause no effects. The last known impact of a massive object, probably over 10 km in diameter, occurred 65 million years ago and caused mass extinctions, including that of the dinosaurs (resulting in the Chicxulub crater mentioned earlier).

On 30 June 1908, an extremely powerful explosion occurred near the Podkamennaya Tunguska River in Russia (Fig. 115). This *Tunguska Event,* as it is called, is believed to have been caused by the air burst of a meteoroid or comet at an altitude of 5 to 10 kilometres above the Earth. The object is estimated to have had a diameter of several tens of metres.

There was no crater as there was no impact as such but the force of the air burst knocked down an estimated 80 million trees over an area greater than 2,000 square kilometres.

These scenarios have been a favourite subject for science fiction writers and film producers (Fig. 116).

Fig. 115 Flattened trees at Tunguska (1908). © NASA

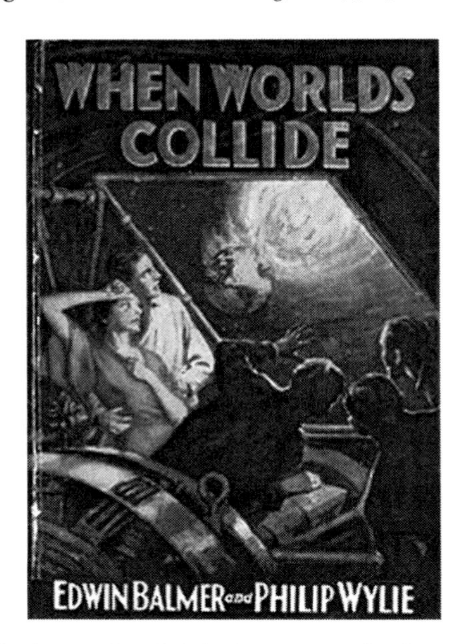

Fig. 116 When Worlds Collide book cover. © Frederick A. Stoker, 1933. Filmed by Paramount Pictures in 1951.

But, could the Earth actually be destroyed as a result of an astronomical impact? Yes it could although it would be an exceedingly rare event (obviously, as it hasn't happened so far), since it would probably require an object several hundreds of miles in diameter to wreak that degree of devastation.

However, one could argue that since the last massive impact was 65 million years ago (although that didn't actually destroy the Earth), perhaps we are due for another one soon. On the positive side, there are now many sophisticated surveillance devices both on Earth and in spacecraft that are continuously monitoring the movements of numerous Near Earth Objects, especially the really large ones. Should such an object be detected and should its orbit bring it on a collision course with Earth, then there would usually be many decades of warning.

Although it's theoretically possible, the chances of suddenly finding a massive Near Earth Object that is calculated to impact in hours or days is extremely remote.

The whole subject of Near Earth Objects and possible collisions with the Earth is now an entire topic of its own. *Spaceguard* is one of several international organisations whose task is to monitor the threat posed by Near Earth Objects. There is even a hazard scale, known as the *Torino Scale* which runs from 0 (no risk) up to 10 (certain collision causing a global catastrophe). Methods for preventing an impending collision, probably by use of space-borne high power lasers or nuclear explosions, are being considered and researched by the relevant authorities.

An astronomical impact with a sufficiently massive asteroid, comet, or meteoroid could end life on Earth. It's possible, but extremely unlikely.

Gamma Ray Burst

A gamma ray burst is a flash of extremely powerful gamma rays emitted from a super nova (exploding star). The first one was detected by a spy satellite in 1967. All the ones that have been observed so far have originated in distant galaxies billions of light years away.

The fact that the radiation has travelled so far and is still easily detectable on Earth gives some impression as to how powerful it really is.

It's been estimated that a single gamma ray burst releases as much energy in a few seconds as our Sun will do in 10 billion years. It's also been estimated that gamma ray bursts occur about once in every 100,000 to 1 million years per galaxy.

Fig. 117 Gamma ray burst GRB 090423 © NASA

On 23 April 2009, the American Swift satellite detected a gamma ray burst which turned out to be the most distant, and hence the oldest, object known in the entire Universe (Fig. 117). It was measured as being 13 billion light years away; this means that the light and gamma radiation detected by the satellite started its journey 13 billion years ago. Since the Universe is considered to be 13.7 billion years old, this object comes from an era when the Universe was only about 700 million years old.

No gamma ray bursts have so far been detected as originating in our galaxy, the Milky Way. That's just as well since it's possible that if that were to happen, and the burst was beamed towards Earth, then that could end all life on the planet.

Other natural events

Tsunamis, earthquakes, floods, volcanic eruptions, climate change, over-population, famine and disease can all be of huge proportions and cause massive destruction of property and life.

For example, the Spanish flu outbreak of 1918-1920 was estimated to have killed up to 100 million people worldwide. However terrible, none of these events is likely to threaten the viability of life on Earth, or of the Earth itself.

The death of the Sun

What we've considered so far – wars, astronomical impacts, gamma ray bursts, and other natural events – may happen or they may not. There really is no way of knowing. But, based on detailed studies of other stars, we do know that our Sun has a well-defined life cycle (Fig. 118). In a few billion years, when the Sun's output of radiation increases, its temperature and consequently that on Earth will rise significantly so that liquid water can no longer exist. At that time, all life will become extinct unless we have found ways of dealing with this problem.

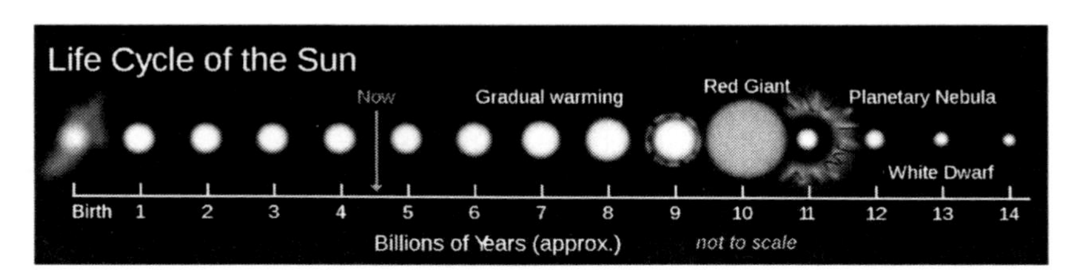

Fig. 118 Life cycle of the Sun with approximate timings (Wikimedia Commons)

As the Sun continues to become hotter over the next few billion years it will also get larger. In about 5½ billion years, the Sun will become a red giant and many times larger than it is now, and it will swallow up Mercury and Venus, and perhaps even the Earth as well.

The timing of these events is necessarily only an estimate. However, the eventual outcome is pretty certain. Life on Earth, and Earth itself, will be destroyed.

That sounds bad. However, a consequence of the Sun getting hotter is that more heat is available to and will reach the outer planets and their moons.

Some of these moons, such as Europa, a moon of Jupiter, and Titan, a moon of Saturn, amongst others, have been suggested as possible sites for the emergence of life.

It's conceivable therefore that in a few billion years or so, when the Sun is getting too hot for us here on Earth, the entire population will have migrated outwards to one or more of these moons. The extra heat given out by the Sun will make the surface temperature much more agreeable on these moons than it is now. That should give Mankind a few more billion years of existence before the Sun finally becomes a white dwarf and, over many more billions of years, gets colder and eventually becomes unable to support any sort of life within its solar system. By that time though, we should have worked out how to deal with this, probably by moving on to another solar system.

The Big Crunch

If the Big Bang model of how the Universe began is accepted, then it could be argued that, in time, the expansion of the Universe will stop and it will start collapsing. As it's been expanding for nearly 14 billion years, it's a fair bet that the Big Crunch won't happen for at least this amount of time, even if it starts now.

It's an interesting concept because, if true, we could then have a new Big Bang to start everything over again, and so on forever more. It's known as the *oscillating Universe.*

We can then make an enormous leap into the far distant future, many trillions and quadrillions of years away. Given enough time, all random events will happen. All possible bridge hands will have been dealt, coins will have come up heads a million times in a row, all the molecules of air in a room will travel in the same direction at the same time and someone standing at the wrong end of the room will suffocate, etc.

George Gamow, in his excellent book *One Two Three...Infinity* (1953, but still available from Amazon and well worth a read) calculated the waiting time for this to happen in an average size room. This came out at $10^{299,999,999,999,999,999,999,999,998}$ seconds. It's impossible to conceive how long this really is but perhaps comparing it with the current age of the Universe of about 10^{17} seconds might put it into some sort of perspective. However, the point is not that this would take an almost unbelievably long time to happen, which we could have guessed, but that given *enough* time, it actually *will* happen.

So imagine this. Given enough time, in a far distant future Universe, life just like ours will develop on a planet just as it did here on Earth. One possible outcome out of the vast number of possible outcomes would be us and our families and friends.

So with an oscillating Universe we could all be re-born one day to live our lives again. It's an interesting, and perhaps to some, a comforting thought.

THE EXPANDING UNIVERSE

It's also possible that the Universe will keep on expanding at a faster and faster rate. In other words, all the galaxies will continue to recede from us at a gradually increasing speed. Eventually, the recession speed will be so fast that the light from these galaxies will never be able to reach us. This is known as the *event horizon*, a barrier beyond which we can never see. Given enough time, everything will have passed the event horizon leaving just our own Milky Way for us to view.

Of course, the same situation will also apply to all the other galaxies in the Universe. Any alien civilisations would, like us, only be able to see their own particular galaxy, all the others including ours having passed *their* event horizon.

THE DEATH OF THE UNIVERSE

If the Universe does continue to expand what will be the eventual outcome of this continued expansion. Notwithstanding the passage of galaxies through event horizons making them unseeable to all other galaxies, will things just continue as they are forever with stars going through their life cycles and finally running out of fuel and exploding as super-novae with new stars and planets then being re-formed from the debris, and so on forever more?

The answer to this is *No* and to understand this we need to consider a principle in physics known as *The Second Law of Thermodynamics*. This is not as abstruse as it sounds since the principle is well-known to everyone.

Imagine a glass of water with some ice in it on a dining room table (Fig. 119). Now imagine the same glass 15 minutes later (Fig. 120).

Fig. 119 Glass of water with ice cubes

Fig. 120 Same glass 15 minutes later

What's happened? The ice has melted. Why? Because the temperature in the room is warmer than the temperature of the ice and so the ice melts because heat always flows from a warmer to a colder place and never the other way around. Something that is hot has more energy than something which is cold and events always happen in a way that tries to equalise the energy distribution.

Eventually, the temperature of the water in the glass will be the same as that of the air in the room (which will have decreased a bit due to the energy taken out of the air to melt the ice). The energy between the water and the room air will then have been equalised.

In simple terms, that's the Second Law of Thermodynamics – left to its own devices, heat always flows from a warmer to a colder place until the temperatures have been equalised. (And yes, there is also a First Law and a Third Law).

This is related to the concept of *entropy*. Entropy is an expression of the state of disorder or randomness of a system. The greater the disorder, the higher the entropy and vice versa.

Nature tends towards maximum entropy for any system left to its own devices. This is a scientific way of saying that the natural tendency is for things to go from order to disorder; ice has an ordered crystalline structure whereas liquid water does not.

If you carefully build a house of cards (Fig. 121) the chances are that it will have collapsed after a few seconds (Fig. 122). Its disorder, or entropy, has increased.

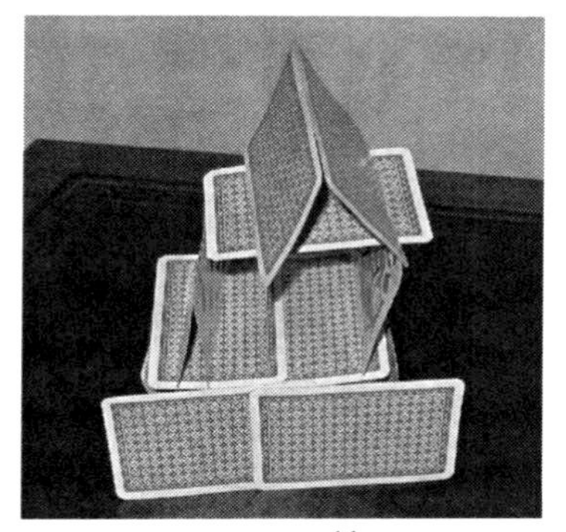

Fig. 121 card house

Fig. 122 5 seconds later

A new box of sugar cubes will have them arranged neatly inside the box. If you empty some of the cubes onto a table, the chances are that they will form a random pile (Fig. 123) rather than an ordered structure such as a wall (Fig. 124). The sugar cubes start in a state of low entropy since they are neatly arranged in the box. But when they are tossed out, it's much more likely that the final state will be one of high entropy (randomness) rather than low entropy (order).

Fig. 123 Random pile of sugar cubes

Fig. 124 Ordered pile of sugar cubes

These are just two more examples of increasing disorder – entropy – in a system left to its own devices. As mentioned above, in the case of the glass of water (Figs. 119 and 120) there is more disorder, entropy, in the glass with the melted ice since solid ice has a crystalline ordered structure whereas liquid water does not.

Now back to the Universe. Given enough time, the Universe will eventually reach a state of maximum entropy, that is, all the energy will have been equally distributed throughout the entire Universe and everywhere will be the same temperature and pressure and everything will be randomly arranged with no structure. This means that no chemical or physical reactions can take place; nothing happens and everything just stops. The Universe has died.

But don't worry too much. This state of affairs will be a long time coming if it comes at all. Calculations from some of Professor Stephen Hawking's work give a lower limit to this of 10^{100} years. Compare this to the current age of the Universe of a little more than 10^{10} years.

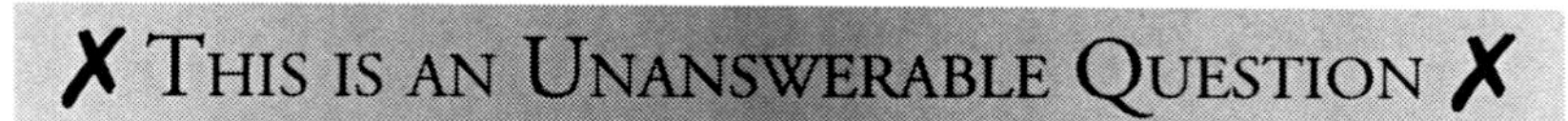
✗ This is an Unanswerable Question ✗

BEST GUESS ANSWER

Life on Earth, and Earth itself, do face realistic threats to their existence. Life could be seriously compromised and even annihilated as a result of nuclear, chemical or biological wars. We have to hope though that no-one would be that stupid. So far, even though these weapons have been available for decades, we have escaped, and we have to hope that our Governments will continue to keep us safe.

Natural climactic and infectious disasters also pose a threat, but none are likely to be of sufficient magnitude to actually wipe out all life on Earth or destroy the Earth itself.

There could however be truly devastating consequences from an astronomical impact or from a gamma ray burst. Fortunately these are extremely rare events which are closely monitored and though it's possible that we could succumb to such a thing, one has to say that it is very unlikely.

Life on Earth, and the Earth, will eventually come to an end as our star, the Sun, enters the later stages of its life cycle. These events are billions of years into the future, and by then we should have developed ways of emigrating to one or more of Jupiter and/or Saturn's moons, which could provide a home for a few more billion years.

After that, when our Sun is a white dwarf, we'd have to go a bit further afield to a new solar system. That seems like quite a journey for a few billion people but in a few billion years, who knows what we'll be able to do?

Ultimately though, current thinking is that one day, many multiple trillions of years into the future, the Universe itself will come to an end, perhaps through a Big Crunch event or by achieving a state of maximum entropy so that all processes just stop. Whether a new Universe would form, or whether there are already vast numbers of other Universes in existence (Question 2) are unanswerable questions.

Question 18:

Will we ever know everything?

Imagine you're a fish swimming about near a beach. You have eyes so maybe you perceive some sort of image of this dry stretch of land that has lots of creatures lying down on it (Fig. 125). But what you certainly aren't aware of are the digital cameras and ipods and mobile phones in these creatures' hands. Although these items may only be 400 metres away, they are separated from you by more than mere distance; they are separated by 400 million years of evolution.

Fig. 125 Possible fishes' view of a beach

So are there things in the Universe that are separated from Man in a similar way? Are there things out there that we do not have the capacity to understand? And are there things still to be invented that we can't even dream about now?

These are hard questions to answer either scientifically or philosophically. We certainly know a great deal more than we did even 10 years ago, and the sum of our knowledge is increasing ever more rapidly due to better and more powerful equipment.

It's probably safe to say that, in certain limited areas, we may well eventually know everything. For example, the field of particle physics is well-funded and is a busy area of research. It's entirely possible that in perhaps a hundred years we will know everything about the fundamental particles of our Universe. Similarly with medicine and genetics, both of which are also well-funded and well-researched, it's a fair guess that in a similar time scale we will know everything about all diseases and how to treat them.

These however are just linear projections from current knowledge. Some things are predictable in this way even though their implementation may require new technology. For example, an ancient Egyptian boat builder living 5,000 years ago might have been watching birds flying gracefully overhead as he and his colleagues were at work, and pondered whether Man might ever be able to fly (Fig. 126). His idea was a linear progression from what he saw even though it took until the early 19th Century for Sir George Cayley to develop a glider that could carry a human passenger.

It's easy enough to predict developments of existing technologies, and the step from bird flight to manned flight is really in this category. But what are hard to predict are new concepts.

Our ancient Egyptian wouldn't have been wondering about mobile telephones, digital cameras, radio and television, computers and the internet, because these were not predictable from contemporary knowledge and experience.

Fig. 126 Ancient Egyptian boat builders and some birds (modified from an illustration in Ancient Records of Egypt by Bearsted, 1922)

In the same way, the future may bring us entirely new concepts that we cannot imagine now. And of course, when and if that future comes, the same could then be said about *that* future's future. So on that basis, no – we will never know everything.

Finally, just like the fish swimming near the beach do not have the capacity to understand, say, a mobile telephone, is it possible that we do not have the capacity to understand certain things that may be around us? It would be rather arrogant of us to deny this, since some aspects of the design and function of the human body could be improved quite significantly.

Here are just a few ideas.

EYES THAT ARE SENSITIVE TO THE ENTIRE ELECTROMAGNETIC SPECTRUM

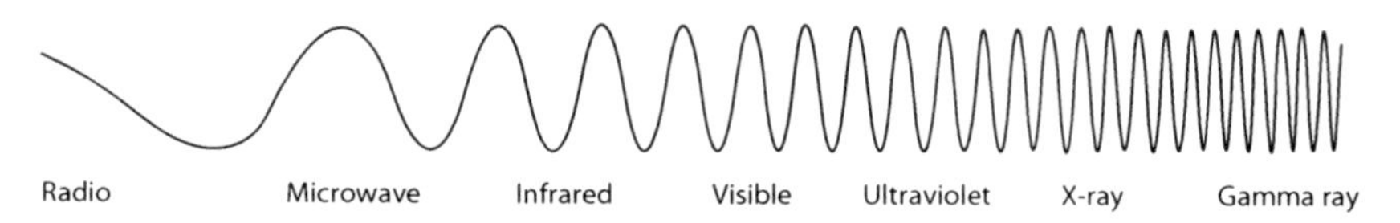

Fig. 127 The electromagnetic spectrum

The electromagnetic spectrum is the entire rage of wavelengths from long radio waves down to short gamma rays. Although there is a little variation in the eyes of different creatures, generally eyes are sensitive only to a relatively narrow range of wavelengths known, obviously, as the *visible spectrum*, as indicated in Fig. 127. Seeing the entire spectrum like this makes it clear how much we *can't* see.

But what if our eyes could perceive other wavelengths? What else could we see? Inside solid objects like Superman with his X-ray vision perhaps?

Superman was always drawn as if he was beaming X-rays onto his subjects as if he had an X-ray generator in his head (Fig. 128). However, this is not how X-ray machines work. The X-ray image is captured on film or other sensitive material *after* the rays have passed through the subject. It makes much more sense to imagine that Superman had a retina that was sensitive to X-rays. If that were the case, then any background radiation of X-rays that fell onto his subjects would be deflected according to the solidity of the subject, and, like an airport luggage scanner or medical X-ray, his eyes, if they were sensitive enough, could then detect items hidden from view to ordinary humans.

Surprisingly, there are quite a lot of articles on the internet as to how Superman's X-ray vision could work.

Fig. 128 Superman and his X-ray vision © DC Ccomics

What about radio waves? We can get some idea of what seeing radio waves could mean by comparing images from optical and radio telescopes (Fig. 129).

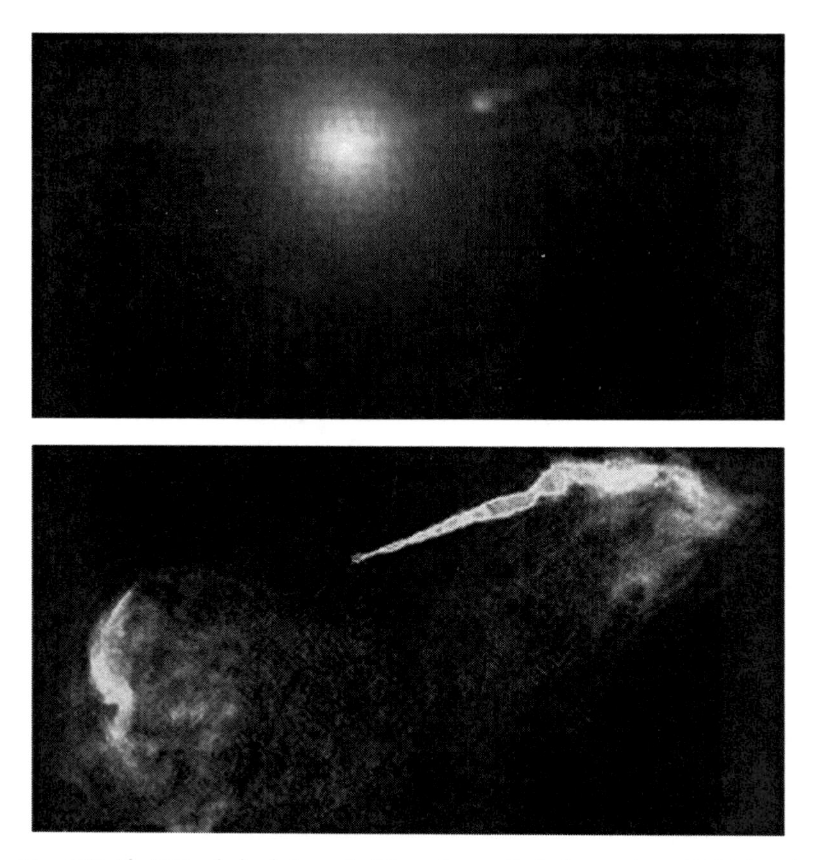

Fig. 129 The M87 nebula photographed through an optical telescope (top - © NASA) and through a radio telescope (bottom - image courtesy of NRAO/AUI/NSF)

Look at all the extra detail in the radio image.

It is clear therefore that we are only seeing a very small part of the World whilst our eyes are only sensitive to quite a narrow range of wavelengths. We have become used to dealing with this, and it serves our purposes, but how much better might it be if we could 'see' all the other wavelengths?

EYES WITH ZOOM LENSES

Anyone familiar with modern cameras will know that zoom lenses act like telescopes and bring distant objects closer. They achieve this by altering the focal length of the lens, and this requires a complex arrangement of separate lenses that move relative to one another during the zooming process. Human and animal eyes do not have this capability.

But what if they did? Imagine trying to hit a target with a bow and arrow and then imagine how much easier it would be if you could zoom in on the target (Fig. 130).

Or how much more useful it would be to be able to spot prey at a greater distance.

Fig. 130 Zooming in on a target

Photographic memory

The ability to recall information in extreme detail after a brief exposure is usually referred to as *photographic memory;* the technical name is *eidetic memory*. Examples would include the ability to recall the exact design of a large crossword and the names and positions of dozens of objects on a tray. This ability should be distinguished from feats such as learning the order of a shuffled deck of cards, since these almost always rely on a system of mnemonics.

Provided that one could choose when to bring this memory into play, it would clearly be an extremely useful skill that is currently enjoyed by only very few people.

The ability to biosynthesise vitamins

Vitamins are substances that are needed in small quantities by organisms to grow and remain healthy. Different organisms have different requirements; humans need 13 vitamins (A; B1; B2; B3; B5; B6; B7; B9; B12; C; D; E; and K). Implicit in this definition is the fact that the human body cannot make these substances and therefore has to obtain them from its diet.

A sufficient supply of vitamins is essential for good health, and a shortage can lead to illness and death. Well-known examples are vitamin C (ascorbic acid) and its deficiency disease *scurvy*, and vitamin D with its deficiency disease *rickets*.

Interestingly, guinea pigs are like humans in this respect since unlike most other small animals, they also cannot produce their own vitamin C so need to have it added to their diets.

Although most people with an adequate diet are very unlikely to become deficient in any vitamins, this is far from the case in poorer nations where deficiency diseases are rife. But if the human body could make (*biosynthesise*- produce by the body) all the vitamins it needs, then there would be no deficiency diseases. Somewhere along the evolutionary path we have lost our ability to produce these substances ourselves.

Immunity to infectious diseases

Apart from war, infectious diseases are the most significant cause of death throughout the World, with the poorer nations suffering the most due to lack of adequate hygiene and medication. For example, the influenza pandemic of 1918 killed between 25 and 50 million people worldwide and even today influenza kills about half a million people each year.

Animals and humans can be given immunity to certain infectious diseases by *vaccination* (from the Latin *vacca* – cow, since the first vaccine was derived from a cow).

This is a very effective process, so much so that the deadly disease of smallpox was officially declared as eradicated on 9 December 1979 and endorsed by the World Health assembly on 8 May 1980.

If humans and animals were all immune to deadly infectious diseases as a matter of design rather than as a result of inoculation programs, then many millions of lives would be saved.

Obviously, there are many other improvements that one could think of, but the point is to show that at our present stage of development, humans are not perfect creatures in every way. If these imperfections extend to our brain power, then maybe there are things that, like the fish, we just cannot understand.

So will we ever know everything? Even if one day we think that we do, how will we know that it's *everything*?

✗ THIS IS AN UNANSWERABLE QUESTION ✗

BEST GUESS ANSWER

Although we can have some confidence in what might happen in the near future, it would be supremely arrogant to imagine that we could predict events into the far distant future. Some of these events could even be beyond our current understanding, making it impossible to conceive what they might be.

It seems unlikely that we will ever know everything, but in a few million or billion years, who knows? As of now, we really can't say.

And finally...

It is hoped that you have found this to be a stimulating and interesting book. Some readers may wish to delve a little deeper into the topics that have been covered. Wikipedia is a good free source. Although some of its articles may be biased, this is less likely to be a problem with scientific data, and in any case, information can always be checked through alternative reference sources.

Far more detailed information can be found in books on specific topics, including those by Richard Dawkins, Stephen Hawking, Brian Cox and Sir Martin Rees amongst many others.

All the information given in this book is intended to be accurate and up to date. Where *Best Guess* answers have been given, these are based on the opinions of the author.

If there's one message that the author would like to impart it's that people should not make decisions based on wishful thinking but should look at the available evidence and use that to come to a decision just like is done in courts of law. Yes, it would indeed be a lovely idea if we never really die but just move into another type of existence where all our loved ones are waiting for us. Or that there's an all-powerful all-knowing superhuman entity in the sky who looks after us if we behave ourselves and pray to him. Or that some people can make contact with the dead. And so on.

Maybe these things will turn out to be true and maybe they won't. One day we'll probably find out for sure. But meanwhile, don't just blindly accept fantastic ideas because you'd like them to be so. Think about alternative explanations that could be much more likely.

Sir Arthur Conan Doyle thought that cardboard cut-outs of fairies propped up in a garden by hat pins were actually real fairies because he *wanted* them to be real. Nigerian crooks keep sending emails to people about giving them millions of dollars in return for their bank details and a small 'administration' charge, and there are always some people who take up the offer because they want it to be real.

Or how about this, which is an actual email received by the author and presumably many other people as well?

Tax Refund Notification

After the last annual calculation of your fiscal activity, we have determined that you are eligible to receive a tax refund of 965.80 GBP. Please submit the tax refund request and allow us 5-7 days in order to process it.

Click the "Refund Me Now" link below and follow the on screen step in order to have us process your request.

Refund Me Now

Note : You will need to provide a valid bank account in which the funds will be payed to. A refund can be delayed for some reasons, for example submitting invalid records or applying after deadline.

HM Revenue & Customs wish you merry christmas and happy new year in advance

Best Regards,
HM Revenue & Customs

It seems much more reasonable than the Nigerian scam but there are several clues to its falsity such as the use of GBP rather than the £ sign; the need to actually apply for the refund; the spelling and grammatical mistakes; and the fact that there is a deadline for the refund.

People who respond to such a message do so because they want it to be true, and either ignore or just don't even see the evidence that says it's not true.

Comments on the book, and suggestions for improvements or for additional questions to pose, would be gratefully received.

INDEX

A

B

C

D

E

F

G

H

I

J

K

L

M

N

O

P

Q

R

S

T

U

X, Y, Z

Lightning Source UK Ltd.
Milton Keynes UK

177605UK00001B/70/P

9 781908 06256